NOMINA ANATOMICA

FIFTH EDITION

NOMINA ANATOMICA

FIFTH EDITION

*Approved by the Eleventh International Congress of Anatomists
at Mexico City, 1980, together with*

NOMINA HISTOLOGICA
SECOND EDITION

AND

NOMINA EMBRYOLOGICA
SECOND EDITION

*Prepared by
Subcommittees of the
International Anatomical Nomenclature Committee*

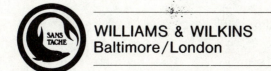

WILLIAMS & WILKINS
Baltimore/London

Made in the United States of America

Library of Congress Cataloging in Publication Data

Main entry under title:

Nomina anatomica.

   1. Anatomy—Nomenclature. 2. Histology—Nomenclature. 3. Embryology—Nomenclature. 4. Veterinary anatomy—Nomenclature. I. International Anatomical Nomenclature Committee. II. International Congress of Anatomists. [DNLM: 1. Anatomy—Nomenclature. 2. Embryology—Nomenclature. 3. Histology—Nomenclature. QS 15 I61n]

QL803.5.N65 1983   611'.0014   82-23894
ISBN 0-683-06550-5

Composed and printed at the
Waverly Press, Inc.
Mt. Royal and Guilford Aves.
Baltimore, MD 21202, U.S.A.

NOMINA ANATOMICA

A Revision by the
International Anatomical Nomenclature Committee
approved by the
Eleventh International Congress of Anatomists
in Mexico City, 1980

PREVIOUS EDITIONS

NOMINA ANATOMICA, 1955 Privately circulated. Printed by Spottiswoode, Ballantyne & Co., Ltd., London.

NOMINA ANATOMICA, 1961 Second Edition. Excerpta Medica, Amsterdam.

NOMINA ANATOMICA, 1963 Second Edition Reprint. Excerpta Medica, Amsterdam.

NOMINA ANATOMICA, 1966 Third Edition. Excerpta Medica, Amsterdam.

NOMINA ANATOMICA, 1968 Third Edition Reprint with Index. Excerpta Medica, Amsterdam.

NOMINA ANATOMICA, 1977 Fourth Edition, with NOMINA HISTOLOGICA and NOMINA EMBRYOLOGICA. Excerpta Medica, Amsterdam.

CONTENTS

CONTENTS

CONTENTS

A vi

CONTENTS

A viii

CONTENTS

SYSTEMA NERVOSUM, 63

CONTENTS

STYLE USAGE

1. Terms within squared brackets are officially recognized synonyms or alternatives.
   Examples:
   > Arteria dorsalis nasi [Arteria nasi externa]
   > Sulci arteriosi [arteriales]
   > Cavitas [Regio] abdominalis

   The second and third examples illustrate an alternative for a part only of the full term.
2. Terms for some structures which are bilateral, and therefore must be qualified as right (dexter) or left (sinister), are printed as follows:
   > Pulmo dexter/sinister

   Similarly, terms for comparable structures in the male and female are printed thus:
   > Nervi scrotales/labiales posteriores
3. Terms are placed in rounded brackets for several purposes.
   Firstly, to indicate that the structure named is inconstant. Examples:
   > (Ossa suturalia)
   > (Sutura frontalis [Sutura metopica])

   Secondly, to indicate certain unofficial but important alternatives. Examples:
   > Tractus corticospinalis (T. pyramidalis)
   > Nucleus nervi facialis (Nucleus facialis)

   Thirdly, to indicate additional components of certain terms which are frequently omitted. Examples:
   > Splenium (corporis callosi)
   > Arachnoidea (mater)
4. Italic type was used for ontogenetic terms in the Third Edition of *Nomina Anatomica*. Most of these terms have been deleted, since they occur in *Nomina Embryologica*. Those which remain are set in italic type.

   Italic type is also now used for subsidiary headings, since the printers preferred to avoid the 'bold' type which was used extensively in the Third Edition.

BRIEF CHRONOLOGY

1895   The Basle *Nomina Anatomica* (B.N.A.) in Latin.

1933   The Birmingham Revision of B.N.A. (B.R.) in Latin and English.

1936   The Jena *Nomina Anatomica* (J.N.A.) in Latin.

1950   Fifth International Congress of Anatomists at Oxford: decision to institute a new body, the International Anatomical Nomenclature Committee (I.A.N.C.).

1952   Preliminary Meeting (supported by C.I.O.M.S.) in London (CIBA Foundation): Confined to general principles. Honorary Secretary, T. B. Johnston.

1954   London Meeting hosted by Ciba Foundation, supported by UNESCO: preliminary lists circulated by T. B. Johnston and G. A. G. Mitchell.

1955   Sixth International Congress of Anatomists at Paris: *Nomina Anatomica*, First Edition, approved and subsequently published. Separate subcommittees set up and rules promulgated. G. A. G. Mitchell became Honorary Secretary.

1960   Seventh International Congress of Anatomists at New York: further revision approved. Subcommittees for Histology and Embryology set up.

1961   *Nomina Anatomica*, Second Edition, edited by G. A. G. Mitchell, published.

1963   Reprint of the Second Edition published (mistakenly described as Third Edition).

1965   Eighth International Congress of Anatomists in Wiesbaden: further amendments. Finance Subcommittee set up.

1966   *Nomina Anatomica*, Third Edition, edited by G. A. G. Mitchell, privately published.

1968   Embryology Subcommittee met in London (Ciba Foundation) to finalize a provisional list.

1969   Histology Subcommitee met in Moscow (Ministry of Public Health of U.S.S.R.).

1970   Ninth International Congress of Anatomists in Leningrad: provisional lists of *Nomina Histologica* and *Nomina Embryologica* distributed gratis. I.A.N.C. meeting attended by representatives of International Committee on Veterinary Anatomical Nomenclature and of International Committee for Avian Anatomical Nomenclature with a view to standardization. G. A. G. Mitchell succeeded W. Bargmann as Chairman; R. Warwick succeeded G. A. G. Mitchell as Honorary Secretary. Combined volume of *Nomina Anatomica*, *Histologica*, *et Embryologica* approved.

1975   Tenth International Congress of Anatomists in Tokyo: revisions of all three lists presented and approved, but magnitude of changes and corrections delayed publication. L. B. Arey succeeded G. A. G. Mitchell as Chairman. R. T. Woodburne elected Vice-Chairman.

1977   *Nomina Anatomica*, Fourth Edition, edited by R. Warwick with *Nomina Histologica*, edited by T. E. Hunt and *Nomina Embryologica*, edited by L. B. Arey and D. Rudnick, published by Excerpta-Medica, Amsterdam/Oxford.

1980   Eleventh International Congress of Anatomists in Mexico: Further revision of all three lists discussed and recommendations approved. Meetings held in parallel with those of International Committee on Veterinary Anatomical Nomenclature. R. T. Woodburne succeeded L. B. Arey as Chairman.

The above events are described in full in previous editions, and especially the fourth, together with the names of the very numerous individuals involved.

INTRODUCTION TO FIFTH EDITION

All previous editions of *Nomina Anatomica* have included an historical account of the various attempts to standardize anatomical nomenclature and, in particular, the work of the International Anatomical Nomenclature Committee (I.A.N.C.). This account was naturally accumulative, and in order to save space and to keep the cost of this volume as low as possible, only a very brief chronology of events has been included. For the same reasons no indices have been added; terms are arranged systematically and, within each system, regionally. This makes the search for particular terms relatively simple. Moreover, there are detailed tables of contents for each major section. Details of membership also have been severely abbreviated and are limited to the needs of the general reader. Everything has in fact been done to reduce the cost of this edition and thus to make it more widely available.

As editor I must accept all the blame for the inadequacies of this revision. I have tried to amend a large number of errors and to carry out, as democratically as possible, all the changes finally approved in Mexico in 1980; but I am regretfully aware that the result is not entirely satisfactory. However, it is essential that this revision should now appear, and for the delay in this I apologize. The next opportunity for amendment and addition of new terms will be at the 12th International Congress of Anatomists in London, 1985. This volume will at least provide the working basis for the next revision. With this venue and its date in mind anyone who wishes to criticize, comment, or suggest any changes must do so in good time, preferably at least six months before the Congress. The delay in publishing this volume is, in part, due to the late arrival of extensive revisional lists. These can only be effectively considered at regional anatomical congresses at which an adequate and appropriate group of I.A.N.C. members can be convened, and these are inevitably infrequent occasions. However, comments of any kind may at any time be addressed to me by individuals or to the chairman of national nomenclatural committees. Alternatively, they may be sent directly to the Conveners or Secretaries of the appropriate Subcommittees, whose names and addresses appear in the section on Membership.

Our financial support has never been wholly adequate, despite generous help in the past from UNESCO, the Royal Society of London, and a minority of national anatomical societies. It would be invidious to name the latter, though much easier than to list those who appear to overlook our inevitable expenses. We are, for example, unable to afford regular secretarial help, and it is not always possible to press extra work upon busy departmental secretaries. In this connection I must express my gratitude to my wife, Carolyn, for much help of this kind.

In financial matters the I.A.N.C. has never been served better than by Professor Raymund Zwemer, who became our Honorary Treasurer in 1965. As a member of the State Department in Washington, UNESCO, the Federation of American Societies for Experimental Biology (FASEB), and of many other such organizations, he was able to secure considerable practical and sometimes financial aid for the I.A.N.C. We are, through him, much indebted in particular to FASEB. Unfortunately for us, and for his family, to whom I have already extended the sympathy of all his colleagues in the I.A.N.C., his very full and active life ended late in 1981. To no one do we owe more in furthering the aims of the I.A.N.C., and I should like to dedicate this edition of *Nomina Anatomica* to him.

ROGER WARWICK
Honorary Secretary
International Anatomical Nomenclature Committee,
℅ The Department of Anatomy, Guy's Hospital Medical School,
London SE1 9RT, Great Britain

MEMBERSHIP

Many changes in membership have occurred since publication of the Fourth Edition, a few by death, many more by resignation and replacement. To all, for their individual services to the International Anatomical Nomenclature Committee (I.A.N.C.), the world of anatomists should accord its appreciation. The deaths of Professor M. J. Hogan, C-H. Hjortsjo, and J. J. Pritchard are noted here with much regret. Of those who have resigned, two men must be specially mentioned: Professor T. E. Hunt and Professor L. B. Arey. The former stepped into the leadership of the Histology Subcommittee at a time when the early efforts of Japanese and Russian histologists were in danger of foundering, and he was the Convener, architect, and Editor of the first-published *Nomina Histologica* in 1977. Professor Arey has been the Convener of the Subcommittee for Embryology since 1960, and I understand he still wishes to serve in this capacity; but we are losing him as our highly experienced Chairman.

Other changes are too numerous to detail here, but it is pleasing to state that we now have members representing Australia, China, Indonesia, Jordan, Portugal, and Turkey, bringing our total of countries with anatomists involved in I.A.N.C. affairs to 30.

The Subcommittee for Anthropological and Primate Terminology, noted in the Fourth Edition, has been dissolved. A new group of interested experts is being formed by Professor R. Singer.

We welcome Professor R. Crafts as our new Honorary Treasurer; Professor R. T. Woodburne, as already noted, is now our Chairman.

ADDRESSES OF PRESENT MEMBERS

The following abbreviations are used to denote membership of particular Subcommittees:

A —Angiology	NS —Nervous System
AM —Arthrology and Myology	O —Osteology
E —Embryology	SO —Sensory Organs
F —Finance	S —Splanchnology
H —Histology	N —New members

PROF. F. D. ALLAN (E)
Department of Anatomy, George Washington University Medical Center, 2300 I Street, N.W., Washington, D.C. 20037, U.S.A.

PROF. MUNIB AL-WIR (N)
Department of Anatomy, School of Medicine, University of Jordan, Amman, Jordan.

PROF. R. M. AMPRINO (E)
Istituto di Anatomia Umana Normale, Università di Bari, Policlinico, 70124 Bari, Italy.

PROF. L. L. ANTUNEZ (NS)
Martin Mendalde No. 1064, Mexico 12, D.F., Mexico.

PROF. L. B. AREY (E Convener)
The Medical School, Northwestern University, 303 East Chicago Avenue, Chicago, Illinois 60611, U.S.A.

DR. M. ARNOLD (S, N)
Department of Anatomy, University of Sydney, Sydney 2006, New South Wales, Australia.

PROF. C. W. ASLING (S, N)
Department of Anatomy, University of California School of Medicine, San Francisco, California 94143, U.S.A.

PROF. A. BAIRATI (H)
Istituto Anatomico, Università di Milano, Via L. Mangiagalli, 31, Milan, Italy.

PROF. B. I. BALINSKY (E)
19 Oban Avenue, Blairgowri, Johannesburg 2001, South Africa.

PROF. R. BARONE (S)
Laboratoire d'Anatomie, Ecole Nationale Vétérinaire de Lyon, Marcy L'Etoile, 69260 Charbonnièves, France.

PROF. J. V. BASMAJIAN (AM)
McMaster University School of Medicine, Chedoke Rehabilitation Centre, Box 590, Hamilton, Ontario, Canada L8N 3L6.

PROF. J. J. BAUMEL (A)
Department of Anatomy, Creighton University School of Medicine, 2500 California Street, Omaha, Nebraska 68178, U.S.A. (International Committee on Avian Anatomical Nomenclature).

PROF. N. BJORKMAN (E)
Kjve Veterinaer-06 Landbonojscholes, Normal Anatomie, Kobenhaven V. Biilowsve 13, Netherlands.

PROF. D. BODIAN (NS)
Department of Anatomy, Johns Hopkins University School of Medicine, 725 North Wolfe Street, Baltimore, Maryland 21205, U.S.A.

PROF. A. P. BONJEAN (O)
Laboratoire d'Anatomie-Chirurgicale, V.E.R. des Sciences Médicales 111, Université de Bordeaux, 146 rue Leo-Saignat, 33076 Bordeaux Cédex, France.

PROF. J. BORODIN (N)
Medizin Institut Novosibirsk, Krasnj Prospekt N. 52, Novosibirsk 91, U.S.S.R.

PROF. J. G. BOSSY (NS, N)
Laboratoire d'Anatomie, Université de Montpellier I, Faculté de Médecine, Avenue Kennedy 30000 Nîmes, France.

PROF. R. P. BUNGE (O Secretary)
Department of Anatomy, Washington University School of Medicine, 660 South Euclid Avenue, St. Louis, Missouri 63110, U.S.A.

PROF. F. CHAMBON (N)
Laboratoire d'Anatomie, Faculté de Médecine, Université de Rennes, Rennes, France.

NOMINA ANATOMICA

PROF. SZE-CHING CHENG (S)
Department of Anatomy, Shanghai First Medical College, Foong Lin Chiao, Shanghai 200032, China.

PROF. I. CHATAIN (NS Secretary)
Departamento de Morfologia, Universidad del Valle, Division de Salud, Cali, Colombia.

PROF. R. CIHAK (AM)
Department of Anatomy, Faculty of Medicine, Charles University, Prague 2, U. Nemocnice 3, Czechoslovakia.

PROF. W. M. COPENHAVER (H)
University of Miami School of Medicine, P.O. Box 520875, Biscayne Annex, Miami, Florida 22152, U.S.A.

PROF. W. M. COWAN (NS)
Department of Anatomy, Washington University School of Medicine, 660 South Euclid Avenue, St. Louis, Missouri 63110, U.S.A.

PROF. R. C. CRAFTS, (F, Convener) (Honorary Treasurer)
3230 Daytona Avenue, Cincinnati, Ohio 45211, U.S.A.

PROF. G. N. CRAWFORD (A, SO)
Department of Anatomy, University of Sheffield, Sheffield 2TN, England.

PROF. A. C. DAS (F)
Department of Anatomy, King George's Medical College, Lucknow, U.P., India.

PROF. A. DELMAS (AM)
Laboratoire d'Anatomie, Académie de Paris, 45 rue des Saints-Pères, Paris 6e, France.

PROF. L. J. A. DIDIO (S)
Department of Anatomy, Medical College of Ohio, Arlington and S. Detroit Avenues, P.O. Box 6190, Toledo, Ohio 43614, U.S.A.

PROF. T. DONÁTH (NS)
Ist Department of Anatomy, Semmelweiss University Medical School, Tuzolto utca 58, 1450 Budapest IX, Hungary.

DR. MARY DYSON (H)
Department of Anatomy, Guy's Hospital Medical School, London, SE1 9RT United Kingdom.

PROF. ALAITTIN ELHAN (NS)
Anatomi Kursusu, Ankara Universitesi Tip Fakultesi, Ankara, Turkey.

PROF. E. A. ERHART (NS Convener)
Departamento de Anatomia, Faculdade de Medicina, Universidade de São Paulo, Caixa postal 2921, São Paulo, Brazil.

PROF. HOWARD E. EVANS (O)
Cornell University, Ithaca, New York 14853, U.S.A. New York State College of Veterinary Medicine.

DR. J. FABER (E)
Hubrecht Laboratory, Uppsalalaan 1, Universiteitscentrum "De Uithof," Utrecht, The Netherlands.

A 4

PROF. J. FAUTREZ (E)
Anatomy Institute, University of Ghent, Ledeganckstraat 35, B-9000 Gnant, Belgium.

PROF. J. FIX (O, NS)
Marshall University, School of Medicine, Huntington, West Virginia 25704, U.S.A.

PROF. S. A. GILMORE (AM)
Dept. of Anatomy, University of Arkansas for Medical Sciences, 4301 West Markham, Little Rock, Arkansas 72205, U.S.A.

PROF. T. W. GLENISTER (E)
Department of Anatomy, Charing Cross Hospital Medical School, Fulham Palace Road, London W6 8RF, England.

PROF. S. GOMEZ-ALVAREZ (N)
Conjes Consejos Nacionallide Ciencias Morfologicas, Apartado Postal 20, 023 Mexico 20, D.F., Mexico.

PROF. M. H. HAST (N)
Dept. of Otolaryngology and Maxillofacial Surgery, Northwestern University Medical School, 303 East Chicago Ave., Chicago, Illinois 60611, U.S.A.

PROF. R. L. HULLINGER (H Secretary, N)
Department of Anatomy, Purdue University School of Veterinary Medicine, Lynn Hall, West Lafayette, Indiana 47907, U.S.A.

PROF. J. HUREAU (S, N)
Laboratoire d'Anatomie, Faculté de Médecine Necker, 156, Rue de Vaugirard (XVe), Paris, France.

PROF. S. IURATO (SO)
Cattedra di Bioacustica, Universitá di Bari, Policlinico, 70124 Bari, Italy.

PROF. M. T. JANSEN (H, N)
Laboratory of Histology and Cell Biology, Nic. Beetstraat 22, 3511 HG Utrecht, The Netherlands

PROF. D. S. JONES (S)
Department of Anatomy, West Virginia University School of Medicine, Morgantown, West Virginia 26506, U.S.A.

PROF. E. N. KEEN (NS)
Department of Anatomy, University of Cape Town Medical School, Observatory 7925, Cape, South Africa.

PROF. M. J. KOERING (H)
Department of Anatomy, George Washington University Medical Center, 2300 First Street, Washington, D.C. 20037, U.S.A.

PROF. B. KONIG, JR. (N)
Escola Paulista de Medicina, Disciplina de Anatomia, Rua Botucatu 720 Vila Clementino 04023, São Paulo, Brazil.

PROF. H.-J. KRETSCHMANN (E, NS)
Medizinische Hochschule Hannover, Institut für Anatomie, 3 Hannover-Kleefeld, Karl-Wiechert-Allee 9, West Germany.

PROF. J. KRMPOTIĆ-NEMANIĆ (M)
Zavod za Anatomiju "Drago Perovic," Medicinskog Fakulteta Sveucilista u Zagrebu, Zagreb, Salata 11, Yugoslavia.

PROF. S. KUBIK (A)
Anatomisches Institut der Universität Zürich, Gloriastrasse 19, 8006 Zürich, Switzerland.

PROF. V. V. KUPRIAYANOV (A)
2nd Medical Institute, G-435 M. Pirogov Prospekt, Moscow, U.S.S.R.

PROF. M. A. MACCONAILL (AM)
Department of Anatomy, University College, Cork, Ireland.

DR. J. MCKENZIE (E)
Department of Developmental Biology, Marischal College, Aberdeen, AB9 IAS, Great Britain.

PROF. Z. MAHRAN (E)
Department of Anatomy, Faculty of Medicine, Ein Shams University, Cairo, Egypt.

PROF. A. H. MARTIN (AM)
Department of Anatomy, University of Western Ontario, London, N6A 5CI Canada.

PROF. G. F. MARTIN (N, S)
Ohio State University, Columbus, Ohio 43210, U.S.A.

PROF. J. L. MARTINEZ (A)
Department of Gross Anatomy, Buenos Aires University Medical School, Paraguay 2155, Buenos Aires, Argentina.

PROF. LATA N. MEHTA (H)
Anatomy Department, Grant Medical College, Byculla, Bombay 400 008, India.

DR. E. R. MEITNER (E, N)
Vorstrand des Histologisch, Embryologisches Instituts, Medizinische Fakultät der Kominsky Universität, Muzeálna 6, 036 01 Martin, Czechoslovakia.

PROF. N. MIZUNA (NS)
Department of Anatomy, Faculty of Medicine, Kyoto University, Yoshida-konecho, Sakyo-ku, Kyoto 606, Japan.

PROF. V. MONESI (H, E, N)
Instituto di Istologia ed Embriologia Generale, Facoltá di Medicina e Chirurgia, Universitá di Roma, Citta Universitaria, 00185 Roma, Italy.

PROF. M. MOSCOVICI (F)
Departamento de Morfologia, Universidade Federal Fluminense, Rua Ernani de Mello 101, Niteroi, Rio de Janeiro, Brazil.

PROF. J. NAKAI (S)
Department of Anatomy, Faculty of Medicine, University of Tokyo, Hongo, Tokyo 113, Japan.

PROF. P. NIEUWENHUIS (H)
Department of Histology, State University, Gruningen, The Netherlands.

PROF. H. NISHIMURA (E)
Department of Anatomy, Faculty of Medicine, Kyoto University, Sakyo-ku, Kyoto, Japan.

PROF. T. M. OELRICH (A Convener)
Department of Anatomy, University of Michigan Medical School, Ann Arbor, Michigan 48104, U.S.A.

PROF. M. OKAMOTO (N)
Department of Anatomy, Kyoto University, Faculty of Medicine Yoshida, Sakyo-ku, Kyoto, Japan.

PROF. J. C. PRATES (O, N)
Departamento de Morphologia, Escola Paulista de Medicina, Rua Botucatu 720, São Paulo, Brazil.

PROF. R. O'RAHILLY (E)
Carnegie Embryology Laboratories, University of California, Davis, California 95616, U.S.A.

PROF. A. J. PALFREY (AM Secretary)
Department of Anatomy, Charing Cross Hospital Medical School, Fulham Palace Road, London W6 8RF, England.

PROF. W. J. PAULE (H)
Department of Anatomy, University of Southern California School of Medicine, 2025 Zonal Avenue, Los Angeles, California 90033, U.S.A.

PROF. B. W. PEARSON (SO)
Department of Otorhinolaryngology, Mayo Clinic, Rochester, Minnesota 55901, U.S.A.

PROF. J. A. ESPERANÇA PINA (N)
Universidade Nova de Lisboa, Praçado Principe Real 26, 1000 Lisbon, Portugal.

PROF. PING-YU (N)
Department of Human Anatomy, Xuzhou Medical College, Xuzhou Jiangsu Province, People's Republic of China.

PROF. P. RABISCHONG (AM, N)
Laboratoire d'Anatomie, Faculté de Médiciné, Universite de Montpellier I, 34060 Montpellier, France.

PROF. M. T. RAKHAWY (H)
Department of Anatomy, Faculty of Medicine, Cairo University, Cairo, Egypt.

PROF. A. Y. RIDEAU (AM, N)
Departement de Cinésiologie Experiméntale, Laboratoire d'Anatomie, 34 Rue du Jardin des Plantes, 86 Poitiers, France.

PROF. J. ROSTGAARD (S)
Department of Anatomy, The University, 1 Universitetsparken, 2100 Copenhagen, Denmark.

DR. D. RUDNICK (E Secretary)
Department of Biology, Albertus Magnus College, New Haven, Connecticut 06511, U.S.A.

PROF. G. SATIUKOWA (H)
Moskauer Mediz Institut "Setschenow," Marx-Prospekt 18, Moscow K-9, U.S.S.R.

PROFESSOR R. SINGER (N)
Department of Anatomy, University of Chicago, 1025 East 57th Street, Chicago, Illinois 60637, U.S.A.

PROF. SOEMAITI AHMAD MUHAMMAD (S, N)
Department of Anatomy, Faculty of Medicine, Gadjah Mada Univeristy, Yogyakarta, Indonesia.

PROF. I. R. TELFORD (H Convener, N)
Uniformed Services University of the Health Sciences, School of Medicine, 4301 Jones Bridge Road, Bethesda, Maryland 20014, U.S.A.

PROF. K. THEILER (E)
Anatomisches Institut der Universität Zurich, Gloriastrasse 19, 8006 Zürich, Switzerland.

PROF. M. TROTTER (O Convener)
Department of Anatomy, Washington University School of Medicine, 660 South Euclid Avenue, St. Louis, Missouri 63110, U.S.A.

PROF. R. WARWICK (Honorary Secretary)
Department of Anatomy, Guy's Hospital Medical School, London SE1 9RT, and 55 Hall Drive, London SE26 6XL, England.

PROF. A. F. WEBER (H, E)
Department of Veterinary Biology, University of Minnesota College of Veterinary Medicine, St. Paul, Minnesota 55161, U.S.A.

PROF. C. WENDELL-SMITH (S, N)
Department of Anatomy, University of Tasmania, Hobart, Tasmania, Australia.

PROF. H. J. L. WERNECK (S, M)
Departamento de Anatomia, Alameda Ásia, 30. P.O. Box 64, 38400 Ukerlândia, MG, Brazil.

PROF. O. WOLKOVA (H, N)
2 Moskauer Mediz Institut "Pirogow", Malaja Pirogowskaja Str. N1, Moscow 48, U.S.S.R.

PROF. WOO JU-KANG (O)
Institute of Vertebrate Palaeontology and Palaeoanthropology, Academia Sinica, P.O. Box 643, Peking (27), China.

PROF. R. T. WOODBURNE (Chairman, S Convener)
Department of Anatomy, University of Michigan Medical School, Ann Arbor, Michigan 48104, U.S.A.

PROF. LI ZHAO-TE (H, E)
Departments of Histology and Embryology, Beijing Medical College, Beijing, People's Republic of China.

A 8

TERMINI GENERALES

Verticalis	Frontalis	Superficialis
Horizontalis	Occipitalis	Profundus
Medianus	Superior	Proximalis
Coronalis	Inferior	Distalis
Sagittalis	Cranialis	Centralis
Dexter	Caudalis	Peripheralis
Sinister	Rostralis	Radialis
Transversalis	Apicalis	Ulnaris
Medialis	Basalis	Fibularis
Intermedius	Basilaris	Tibialis
Lateralis	Medius	Palmaris
Anterior	Transversus	Volaris
Posterior	Longitudinalis	Plantaris
Ventralis	Axialis	Flexor
Dorsalis	Externus	Extensor
	Internus	

REGIONES ET PARTES CORPORIS

The precise anatomical meaning of the term region is not easy to define; all those included here correspond to some area of the *surface* of the body, and may therefore be called *Regiones Superficiales*. Some of these are little more than geometrical abstractions, but most are well-established and useful terms. Definitions of the boundaries of certain regions, e.g., those of the abdomen, exhibit national variations, but since such concepts are chiefly of clinical application and are not used with exactitude, the imposition of precise definitions is not as important as uniformity in the nomenclature of such regions. Wherever greater precision is required clinicians and others usually make actual measurements from skeletal reference points.

(The term *Regio* is also sometimes used for groups of structures within the body, e.g. *regio infratemporalis, regio pterygoidea, regio poplitea*, etc. These are also referred to as *fossa infratemporalis*, etc., and will be found as such in *Nomina Anatomica*. This usage is, of course, illogical, for a *fossa* is created only by the removal of the contents which form the *region*.)

Linea mediana anterior	Linea medioclavicularis[3]
(Linea sternalis)[1]	(Linea mamillaris)[4]
Linea parasternalis[2]	Linea axillaris anterior[5]

[1] A vertical line corresponding to the lateral sternal margin—difficult to construct in view of the variable horizontal dimensions of the sternum.
[2] A vertical line equidistant from the sternal and mid-clavicular lines.
[3] This vertical line is customarily used as one of the boundaries of the abdominal regions. A *linea pararectalis*, following the lateral margin of the *musculus rectus abdominis*, is advocated by a minority of anatomists for this purpose, but it is difficult to define in many people.
[4] This was equated with the *linea medioclavicularis* in the third edition, but the variable position of the mamilla renders this untenable.
[5] These lines correspond, of course, to the *plicae axillares*.

NOMINA ANATOMICA

Linea axillaris media [medio-axillaris]
Linea axillaris posterior[5]
Linea scapularis[6]
Linea paravertebralis[7]
Linea mediana posterior
Planum subcostale[8]
Planum transpyloricum[9]
Planum supracristale[10]
Planum intertuberculare[10]
Planum interspinale[10]
Linea preaxillaris
Linea postaxillaris

REGIONES CAPITIS[11]
Regio frontalis
Regio parietalis
Regio occipitalis
Regio temporalis

REGIONES FACIALES
Regio orbitalis
Regio nasalis
Regio oralis
Regio mentalis
Regio infraorbitalis
Regio buccalis
Regio zygomatica

REGIONES CERVICALES[12]
Regio cervicalis anterior [Trigonum
  cervicale anterius]
  Trigonum submandibulare
  Trigonum caroticum
  Trigonum musculare [omotracheale]
  Trigonum submentale
Regio sternocleidomastoidea
  Fossa supraclavicularis minor

Regio cervicalis lateralis [Trigonum
  cervicle posterius]
  Trigonum omoclaviculare[13]
  Fossa supraclavicularis major[13]
Regio cervicalis posterior [R. nuchalis]

REGIONES PECTORALES
Regio presternalis
Fossa infraclavicularis
Trigonum clavipectorale
Regio pectoralis
Regio mammaria
Regio inframammaria
Regio axillaris
Fossa axillaris

REGIONES ABDOMINALES
Regio hypochondriaca[14]
  [Hypochondrium]
Regio epigastrica [Epigastrium]
Regio lateralis[14]
Regio umbilicalis
Regio inguinalis[14]
Regio pubica [Hypogastrium]

REGIONES DORSALES
Regio vertebralis
Regio lumbalis [lumbaris]
Regio sacralis
Regio scapularis
Regio infrascapularis
Trigonum lumbare

REGIO PERINEALIS
Regio analis
Regio urogenitalis

[6] A vertical line through the *angulum scapulae inferius*.
[7] A longitudinal line corresponding to the transverse vertebral processes and hence chiefly of radiological value. It is sometimes defined as a vertical line midway between the scapular and posterior median lines, in which case the *paravertebral* line, as defined above, is termed the *vertebral* line. The Committee preferred the first definition, familiar to radiologists and others.
[8] This plane is level with the inferior limits of the costal margins, i.e. the tenth costal cartilages.
[9] A horizontal plane midway between the superior margins of the *manubrium sterni* and *symphysis pubis*. It does not usually correspond to the level of the pylorus.
[10] These three planes are at the levels of the summits, tubercles, or anterior superior spines of the iliac crests. The most commonly used in definition of the abdominal planes is the *planum interspinale*. The *planun supracristale* is at the level of the fourth lumber spinous process.
[11] These terms can be united as required, e.g. *regio frontoparietalis, parieto-occipitalis*, etc.
[12] These were termed *regiones colli* in the third edition. Controversy over the precise meanings of *cervix* and *collum* in classical Latin is not pertinent, and classical usage varied. Since almost all structures in the *collum* are qualified as cervical, it is pointless to avoid the use of *cervix* for the neck.
[13] These terms are really synonymous.
[14] The terms are, of course, qualified as *dextra* or *sinistra*.

A 10

REGIONES MEMBRI SUPERIORIS

Regio deltoidea

Brachium
Regio [Facies] brachialis anterior
Regio [Facies] brachialis posterior

Cubitus
Regio [Facies] cubitalis anterior
Regio [Facies] cubitalis posterior
Fossa cubitalis
Sulcus bicipitalis lateralis [radialis]
Sulcus bicipitalis medialis [ulnaris]

Antebrachium
Regio [Facies] antebrachialis anterior
Regio [Facies] antebrachialis posterior
Margo lateralis [radialis]
Margo medialis [ulnaris]

Carpus
Regio carpalis anterior
Regio carpalis posterior

Manus
Dorsum manus
Palma manus
Thenar [Eminentia thenaris]
Hypothenar [Eminentia hypothenaris]
Metacarpus
Digiti
Pollex [Digitus primus (I)]
Index [Digitus secundus (II)]
Digitus medius [tertius (III)]
Digitus annularis [quartus (IV)]
Digitus minimus [quintus (V)]
Facies digitales ventrales [palmares]
Facies digitales dorsales

REGIONES MEMBRI INFERIORIS
Regio glutealis
Sulcus glutealis

Femur [Regio femoralis]
Regio [Facies] femoralis anterior
Trigonum femorale
Regio [Facies] femoralis posterior

Genus
Regio genus [genualis] anterior
Regio genus [genualis] posterior
Fossa poplitea

Crus [Regio cruralis]
Regio [Facies] cruralis anterior
Regio [Facies] cruralis posterior
Sura [Regio suralis]
Regiones talocruralis anterior et posterior

Pes
Regio calcanea [Calx]
Dorsum [Regio dorsalis] pedis
Planta [Regio plantaris] pedis
Margo lateralis [fibularis] pedis
Margo medialis [tibialis] pedis
Tarsus
Metatarsus
Digiti
Hallux [Digitus primus (I)]
Digitus secundus, tertius, quartus (II, III, IV)
Digitus minimus [quintus (V)]

OSTEOLOGIA

Systema skeletale
Pars ossea
Periosteum
Endosteum
Substantia corticalis
Substantia compacta
Substantia spongiosa [trabecularis]
Pars cartilaginosa
Perichondrium
Skeleton axiale
Skeleton appendiculare
Os longum
Os breve
Os planum
Os irregulare
Os pneumaticum
Epiphysis
Diaphysis
Metaphysis
Cartilago epiphysialis
Linea epiphysialis[15]

[15] This is the *plane* or plate which marks the site of recent ossification of an epiphysial cartilage.

NOMINA ANATOMICA

Facies articularis
Cavitas medullaris
Medulla ossium flava
Medulla ossium rubra
Foramen nutricium (nutriens)
Canalis nutricius (nutriens)
Punctum (Centrum) ossificationis
  Primarium
  Secundarium

SKELETON AXIALE

CRANIUM

Cavitas cranii
  Pericranium
  Lamina externa
  Diploë
    Canales diploici
  Lamina interna
  Sulcus sinus sagittalis superioris
  Foveolae granulares
  Impressiones digitatae [gyrorum]
    Sulci venosi
    Sulci arteriosi [arteriales]
  (Ossa suturalia)

(Norma verticalis)[16]
  Bregma
  Vertex
  Occiput

Norma facialis[16]
  Frons
  Nasion
  Gnathion

Orbita
  Aditus orbitae
    Margo orbitalis
      Margo supraorbitalis
      Margo infraorbitalis
      Margo lateralis
      Margo medialis
  Paries superior
  Paries inferior

Paries lateralis
Paries medialis
  Foramen ethmoidale anterius
  Foramen ethmoidale posterius
  Sulcus lacrimalis
  Fossa sacci lacrimalis
Fissura orbitalis superior
Fissura orbitalis inferior

Cavitas nasi
  Septum nasi osseum
  Apertura piriformis [nasalis anterior]
  Meatus nasalis superior
  Meatus nasalis medius
  Meatus nasalis inferior
    Canalis nasolacrimalis
  Recessus spheno-ethmoidalis
  Meatus nasopharyngeus
  Choanae
  Foramen sphenopalatinum

Norma lateralis[16]
  Pterion
  Asterion
  Gonion
  Fossa temporalis
  Arcus zygomaticus
  Fossa infratemporalis
    Fossa pterygopalatina
    Fissura pterygomaxillaris

Norma basilaris [Basis cranii externa]
  Foramen jugulare
  Fissura sphenopetrosa
  Fissura petro-occipitalis
  Foramen lacerum[17]
  Palatum osseum
  Canalis palatinus major
  Foramen palatinum majus
    Fossa incisiva
      Canalis incisivus
      Foramina incisiva
    (Torus palatinus)
  Canalis palatovaginalis
  Canalis vomerovaginalis
  Canalis vomerorostralis[18]

---

[16] *Norma verticalis, Norma facialis,* etc., though not structures, are nevertheless terms frequently found in anatomical texts.
[17] In life this foramen does not exist, being occupied by cartilage.
[18] This canal is between the vomer and sphenoidal rostrum.

A 12

NORMA OCCIPITALIS[16]
  Inion
  Lambda

BASIS CRANII INTERNA
  Fossa cranii anterior
  Fossa cranii media
  Fossa cranii posterior
    Clivus
    Sulcus sinus petrosi inferioris

FONTICULI CRANII
  Fonticulus anterior
  Fonticulus posterior
  Fonticulus sphenoidalis [anterolateralis]
  Fonticulus mastoideus [posterolateralis]

OSSA CRANII [CRANIALIA]

OS OCCIPITALE
Foramen magnum
  Basion
  Opisthion
Pars basilaris
  Tuberculum pharyngeum
Pars lateralis
Squama occipitalis
  Margo mastoideus
  Margo lambdoideus
  (Os interparietale)
Condylus occipitalis
Canalis condylaris
Canalis hypoglossi
Fossa condylaris
Tuberculum jugulare
Incisura jugularis
Processus jugularis
Processus intrajugularis
Protuberantia occipitalis externa
(Crista occipitalis externa)
Linea nuchae suprema
Linea nuchae superior
Linea nuchae inferior
Eminentia cruciformis
Protuberantia occipitalis interna
(Crista occipitalis interna)
Sulcus sinus transversi
Sulcus sinus sigmoidei
Sulcus sinus occipitalis
(Processus paramastoideus)

OS SPHENOIDALE
Corpus
  Jugum sphenoidale
  Sulcus prechiasmaticus
  Sella turcica
    Tuberculum sellae
    (Processus clinoideus medius)
    Fossa hypophysialis
    Dorsum sellae
    Processus clinoideus posterior
  Sulcus caroticus
  Lingula sphenoidalis
  Crista sphenoidalis
  Rostrum sphenoidale
  Sinus sphenoidalis
    Septum sinuum sphenoidalium
    Apertura sinus sphenoidalis
  Concha sphenoidalis

Ala minor
  Canalis opticus
  Processus clinoideus anterior
  Fissura orbitalis superior

Ala major
  Facies cerebralis
  Facies temporalis
  Facies maxillaris
  Facies orbitalis
  Margo zygomaticus
  Margo frontalis
  Margo parietalis
  Margo squamosus
  Crista infratemporalis
  Foramen rotundum
  Foramen ovale
  (Foramen venosum)
  Foramen spinosum
  (Foramen petrosum)
  Spina ossis sphenoidalis
  Sulcus tubae auditivae (auditoriae)

Processus pterygoideus
  Lamina lateralis (processus pterygoidei)
  Lamina medialis (processus pterygoidei)
  Incisura pterygoidea
  Fossa pterygoidea
  Fossa scaphoidea
  Processus vaginalis
    Sulcus palatovaginalis
    Sulcus vomerovaginalis
  Hamulus pterygoideus

Sulcus hamuli pterygoidei
Canalis pterygoideus
Processus pterygospinosus

Os temporale

Pars petrosa
Margo occipitalis
Processus mastoideus
Incisura mastoidea
Sulcus sinus sigmoidei
Sulcus arteriae occipitalis
Foramen mastoideum
Canalis facialis
　Geniculum canalis facialis
Canaliculus chordae tympani
Apex partis petrosae
　Canalis caroticus
　　Canaliculi caroticotympanici
　　Canalis musculotubarius
　　Semicanalis musculi tensoris
　　tympani
　　Semicanalis tubae auditivae
　　Septum canalis musculotubarii
Facies anterior partis petrosae
　Tegmen tympani
　Eminentia arcuata
　Hiatus canalis nervi petrosi majoris
　　Sulcus nervi petrosi majoris
　Hiatus canalis nervi petrosi minoris
　　Sulcus nervi petrosi minoris
　Impressio trigeminalis
Margo superior partis petrosae
　Sulcus sinus petrosi superioris
Facies posterior partis petrosae
　Porus acusticus internus
　Meatus acusticus internus
　Fossa subarcuata
　　Aqueductus vestibuli
　　Apertura externa aqueductus
　　vestibuli
Margo posterior partis petrosae
　Incisura jugularis
　Processus intrajugularis
　Canaliculus cochleae
　Apertura externa canaliculi cochleae
Facies inferior partis petrosae
　Fossa jugularis
　　Canaliculus mastoideus
　　Incisura jugularis
　Processus styloideus
　Foramen stylomastoideum

Canaliculus tympanicus
Fossula petrosa
Cavitas tympanica
(*see* page A81, Organum
　vestibulocochleare)
Fissura petrotympanica
Fissura petrosquamosa
Fissura tympanomastoidea
Fissura tympanosquamosa

Pars tympanica
Annulus [Anulus] tympanicus
　Meatus acusticus externus
　Porus acusticus externus
　Spina tympanica major
　Spina tympanica minor
　Sulcus tympanicus
　Incisura tympanica
　Vagina processus styloidei

Pars squamosa
Margo parietalis
Incisura parietalis
Margo sphenoidalis
Facies temporalis
Sulcus arteriae temporalis mediae
Processus zygomaticus
　Crista supramastoidea
　Foveola suprameatica [suprameatalis]
　(Spina suprameatica [suprameatalis])
Fossa mandibularis
　Facies articularis
Tuberculum articulare
　Facies cerebralis

Os parietale
Facies interna
　Sulcus sinus sigmoidei
Facies externa
　Linea temporalis superior
　Linea temporalis inferior
　Tuber parietale
Margo occipitalis
Margo squamosus
Margo sagittalis
Margo frontalis
Angulus frontalis
Angulus occipitalis
Angulus sphenoidalis
Angulus mastoideus
Foramen parietale

OS FRONTALE
  Squama frontalis
    Facies externa
      Tuber [Eminentia] frontale
      Arcus superciliaris
      Glabella
      Margo supraorbitalis
        Incisura supraorbitalis/Foramen
          supraorbitale
        Incisura frontalis/Foramen
          frontale
    Facies temporalis
      Margo parietalis
      Linea temporalis
      Processus zygomaticus
    Facies interna
      Crista frontalis
      Sulcus sinus sagittalis superioris
      Foramen caecum [cecum]
  Pars nasalis
    Spina nasalis
    Margo nasalis
  Pars orbitalis
    Facies orbitalis
      (Spina trochlearis)
      Fovea trochlearis
      Foramina ethmoidalia
      (*see* page A12, orbita)
      Fossa glandulae lacrimalis
  Incisura ethmoidalis
  Sinus frontalis
    Apertura sinus frontalis
    Septum sinuum frontalium

OS ETHMOIDALE
  Lamina et Foramina cribrosa
  Crista galli
    Ala cristae galli
  Lamina perpendicularis

Labyrinthus ethmoidalis
  Cellulae ethmoidales
  Lamina orbitalis
  Foramina ethmoidalia
  (*see* page A12, orbita)
  (Concha nasalis suprema)
  Concha nasalis superior
  Concha nasalis media
  Bulla ethmoidalis
  Processus uncinatus
  Infundibulum ethmoidale
  Hiatus semilunaris

CONCHA NASALIS INFERIOR
  Processus lacrimalis
  Processus maxillaris
  Processus ethmoidalis

OS LACRIMALE
  Crista lacrimalis posterior
  Hamulus lacrimalis

OS NASALE
  Sulcus ethmoidalis

VOMER
  Ala vomeris
  Sulcus vomeris
  Crista choanalis vomeris
  Pars cuneiformis vomeris

MAXILLA
Corpus maxillae
  Facies orbitalis
    Canalis infraorbitalis
    Sulcus infraorbitalis
  Margo infraorbitalis
  Facies anterior
    Foramen infraorbitale
    Fossa canina
    Incisura nasalis
    Spina nasalis anterior
    Sutura zyomaticomaxillaris
  Facies infratemporalis
    Foramina alveolaria
    Canales alveolares
    Tuber (Eminentia) maxillae
  Facies nasalis
    Sulcus lacrimalis
    Crista conchalis
    Margo lacrimalis
    Hiatus maxillaris
  Sulcus palatinus major
  Sinus maxillaris

Processus frontalis
  Crista lacrimalis anterior
  Incisura lacrimalis
  Crista ethmoidalis

Processus zygomaticus

Processus palatinus
  Crista nasalis

A 15

(Os incisivum)
Canalis incisivus
  (see Cranium, palatum osseum)
(Sutura incisiva)
Spinae palatinae
Sulci palatini

Processus alveolaris
Arcus alveolaris
Alveoli dentales
Septa interalveolaria
  Septa interradicularia
Juga alveolaria
Foramen incisivum
  (*see* Cranium, palatum osseum)

OS PALATINUM
Lamina perpendicularis
Facies nasalis
Facies maxillaris
Incisura sphenopalatina
Sulcus palatinus major
Processus pyramidalis
Canales palatini minores
Crista conchalis
Crista ethmoidalis
Processus orbitalis
Processus sphenoidalis

Lamina horizontalis
Facies nasalis
Facies palatina
Foramina palatina minora
Spina nasalis posterior
Crista nasalis
Crista palatina

OS ZYGOMATICUM
Facies lateralis
Facies temporalis
Facies orbitalis
Processus temporalis
Processus frontalis
  Eminentia orbitalis[19]
(Tuberculum marginale)
Foramen zygomatico-orbitale
Foramen zygomaticofaciale
Foramen zygomaticotemporale

MANDIBULA

Corpus mandibulae
Basis mandibulae
Symphysis mandibulae
Protuberantia mentalis
Tuberculum mentale
Foramen mentale
Linea obliqua
Fossa digastrica
Spina mentalis
Linea mylohyoidea
(Torus mandibularis)
Fovea sublingualis
Fovea submandibularis
Pars alveolaris
  Arcus alveolaris
  Alveoli dentales
  Septa interalveolaria
  Juga alveolaria

Ramus mandibulae
Angulus mandibulae
(Tuberositas masseterica)
(Tuberositas pterygoidea)
Foramen mandibulae
  Lingula mandibulae
  Canalis mandibulae
Sulcus mylohyoideus
Processus coronoideus
Incisura mandibulae
Processus condylaris
  Caput mandibulae
  Collum mandibulae
  Fovea pterygoidea

OSSICULA AUDITUS (*see* page A82)
Stapes
  Caput stapedis
  Crus anterius
  Crus posterius
  Basis stapedis
Incus
  Corpus incudis
  Crus longum
    Processus lenticularis
  Crus breve
Malleus
  Manubrium mallei
  Caput mallei

[19] An eminence present in most *Ossa zygomatica*.

Collum mallei
Processus lateralis
Processus anterior

OS HYOIDEUM
Corpus
Cornu minus
Cornu majus

COLUMNA VERTEBRALIS

Canalis vertebralis
Corpus vertebrae [vertebralis]
Arcus vertebrae [vertebralis]
    Pediculus arcus vertebrae [vertebralis]
    Lamina arcus vertebrae [vertebralis]
Foramen intervertebrale
Incisura vertebralis superior
Incisura vertebralis inferior
Foramen vertebrale
Processus spinosus
Processus transversus
    Processus costalis
Processus articularis [Zygapophysis]
        superior/inferior

VERTEBRAE CERVICALES [CI–CVII]
    Uncus corporis[20]
    Foramen processus transversi
        [F. vertebrarteriale]
    Tuberculum anterius (T. caroticum)
    Tuberculum posterius
    Sulcus nervi spinalis

ATLAS [CI]
    Massa lateralis
        Facies articularis superior
        Facies articularis inferior
    Arcus anterior
        Fovea dentis
        Tuberculum anterius
    Arcus posterior
        Sulcus arteriae vertebralis
        Tuberculum posterius

AXIS [CII]
    Dens
        Apex dentis

Facies articularis anterior
Facies articularis posterior

VERTEBRA PROMINENS [CVII]

VERTEBRAE THORACICAE [TI–TXII]
    Fovea costalis superior
    Fovea costalis inferior
    Fovea costalis processus transversus

VERTEBRAE LUMBALES [LUMBARES] [LI–LV]
    Processus accessorius[21]
    Processus mammillaris[21]

OS SACRUM [SACRALE] (VERTEBRAE
        SACRALES I–V)
    Basis ossis sacri
        Promontorium
        Ala sacralis
        Processus articularis superior
    Pars lateralis
        Facies auricularis
        Tuberositas sacralis
    Facies pelvica
        Lineae transversae
        Foramina intervertebralia
        Foramina sacralia anteriora [pelvica]
    Facies dorsalis
        Crista sacralis mediana
        Foramina sacralia posteriora
        Crista sacralis intermedia
        Crista sacralis lateralis
        Cornu sacrale
        Canalis sacralis
            Hiatus sacralis
    Apex ossis sacri

OS COCCYGIS [COCCYX] [VERTEBRAE
        COCCYGEAE I–IV]
Cornu coccygeum

OSSA THORACIS

COSTAE [I–XII]
    Costae verae [I–VII]
    Costae spuriae [VIII–XII]
        Costae fluitantes [XI–XII]
    Cartilago costalis

---

[20] A hook-like projection on each side of the superior surface of the 3rd to 7th cervical vertebral bodies.
[21] Although listed only with the lumbar vertebrae, these processes occur also in the 10th, 11th, and 12th thoracic vertebrae.

A 17

Os costate [Costa]
  Caput costae
    Facies articularis capitis costae
    Crista capitis costae
    Collum costae
      Crista colli costae
    Corpus costae
      Tuberculum costae
        Facies articularis tuberculi costae
      Angulus costae
      Sulcus costae
  (Costa cervicalis)
  Costa prima (I)
    Tuberculum musculi scaleni anterioris
    Sulcus arteriae subclaviae
    Sulcus venae subclaviae
  Costa secunda (II)
    Tuberositas musculi serrati anterioris

STERNUM
  Manubrium sterni
    Incisura clavicularis
    Incisura jugularis
  Angulus sterni [sternalis]
  Corpus sterni
  Processus xiphoideus
  Incisurae costales
  (Ossa suprasternalia)

COMPAGES THORACIS[22]
  Cavitas thoracis
  Apertura thoracis superior
  Apertura thoracis inferior
  Sulcus pulmonalis
  Arcus costalis
  Spatium intercostale
  Angulus infrasternalis

SKELETON APPENDICULARE

OSSA MEMBRI SUPERIORIS

Cingulum membri superioris [Cingulum pectorale]

SCAPULA
  Facies costalis [anterior]
    Fossa subscapularis

Facies posterior [Dorsum]
  Spina scapulae
  Fossa supraspinata
  Fossa infraspinata
Acromion
  Facies articularis acromii
  Angulus acromialis
Margo medialis
Margo lateralis
Margo superior
  Incisura scapulae
Angulus inferior
Angulus lateralis
Angulus superior
Cavitas glenoidalis
Tuberculum supraglenoidale
Tuberculum infraglenoidale
Collum scapulae
Processus coracoideus

CLAVICULA
Extremitas sternalis
  Facies articularis sternalis
  Impressio ligamenti costoclavicularis
Corpus claviculae
  Sulcus musculi subclavii
Extremitas acromialis
  Facies articularis acromialis
  Tuberculum conoideum
  Linea trapezoidea

Pars libera membri superioris

HUMERUS
  Caput humeri
  Collum anatomicum
  Collum chirurgicum
  Tuberculum majus
  Tuberculum minus
    Sulcus intertubercularis
      Crista tuberculi majoris
      Crista tuberculi minoris
  Corpus humeri
    Facies anterior medialis
      [anteromedialis]
    Facies anterior lateralis [anterolateralis]
    Facies posterior
      Sulcus nervi radialis
    Margo medialis
      Crista supraepicondylaris medialis

---

[22] This term, recently introduced, denotes the thoracic cage.

(Processus supraepicondylaris)
Margo lateralis
  Crista supraepicondylaris lateralis
Tuberositas deltoidea
Condylus humeri
  Capitulum humeri
  Trochlea humeri
  Fossa olecrani
  Fossa coronoidea
  Fossa radialis
Epicondylus medialis
  Sulcus nervi ulnaris
Epicondylus lateralis

RADIUS
  Caput radii
    Fovea articularis
    Circumferentia articularis
  Collum radii
    Corpus radii
    Tuberositas radii
    Facies anterior
    Facies posterior
    Facies lateralis
    Margo interosseus
    Margo anterior
    Margo posterior
  Processus styloideus
  Tuberculum dorsale
  Incisura ulnaris
  Facies articularis carpi [carpalis]

Ulna
  Olecranon
  Processus coronoideus
    Tuberositas ulnae
  Incisura trochlearis
  Incisura radialis
  Corpus ulnae
    Facies anterior
    Facies posterior
    Facies medialis
    Margo interosseus
    Margo anterior
    Margo posterior
    Crista musculi supinatoris
  Caput ulnae
    Circumferentia articularis
    Processus styloideus

OSSA CARPI [CARPALIA]
  (Os centrale)[23]
  Os scaphoideum
    Tuberculum ossis scaphoidei
  Os lunatum
  Os triquetrum
  Os pisiforme
  Os trapezium
    Tuberculum ossis trapezii
  Os trapezoideum
  Os capitatum
  Os hamatum
    Hamulus ossis hamati
  Sulcus carpi

OSSA METACARPI [METACARPALIA] (I–V)
  Basis metacarpalis
  Corpus metacarpalis
  Caput metacarpalis
  Os metacarpale tertium (III)
  Processus styloideus

OSSA DIGITORUM [PHALANGES]
  Phalanx proximalis
  Phalanx media
  Phalanx distalis
    Tuberositas phalangis distalis
  Basis phalangis
  Corpus phalangis
  Caput [Trochlea] phalangis
  Ossa sesamoidea

OSSA MEMBRI INFERIORIS

Cingulum membri inferioris [Cingulum pelvicum]

OS COXAE [PELVICUM]
  Acetabulum
    Limbus (Margo) acetabuli
    Fossa acetabuli
    Incisura acetabuli
    Facies lunata
  Foramen obturatum [obturatorium]

Os ilii [Ilium]
  Corpus ossis ilii
    Sulcus supra-acetabularis

---

[23] There exists a controversy over the status of this carpal element—whether it is truly carpal or merely a sesamoid bone.

Ala ossis ilii
 Linea arcuata
 Crista iliaca
  Labium externum
   Tuberculum iliacum
  Linea intermedia
  Labium internum
  Spina iliaca anterior superior
  Spina iliaca anterior inferior
  Spina iliaca posterior superior
  Spina iliaca posterior inferior
 Fossa iliaca
 Facies glutea
  Linea glutea anterior
  Linea glutea posterior
  Linea glutea inferior
 Facies sacropelvica
  Facies auricularis
 Tuberositas iliaca

Os ischii [*Ischium*]
 Corpus ossis ischii
 Ramus ossis ischii
  Tuber ischiadicum [ischiale]
 Spina ischiadica [ischialis]
 Incisura ischiadica [ischialis] major
 Incisura ischiadica [ischialis] minor

Os pubis [*Pubis*]
 Corpus ossis pubis
  Tuberculum pubicum
  Facies symphysialis
  Crista pubica
 Ramus superior ossis pubis
  Eminentia iliopubica
  Pecten ossis pubis
  Crista obturatoria
  Sulcus obturatorius
  Tuberculum obturatorium anterius
  (Tuberculum obturatorium posterius)
 Ramus inferior ossis pubis

PELVIS
 Cavitas pelvis
 Arcus pubis [pubicus]
  Angulus subpubicus
 Pelvis major
 Pelvis minor
 Linea terminalis
 Apertura pelvis superior
 Apertura pelvis inferior
 Axis pelvis

Diameter conjugata
Diameter transversa
Diameter obliqua
Inclinatio pelvis

Pars libera membrae inferioris

FEMUR [OS FEMORIS]
 Caput ossis femoris
  Fovea capitis ossis femoris
 Collum ossis femoris
 Trochanter major
  Fossa trochanterica
 Trochanter minor
 (Trochanter tertius)
 Linea intertrochanterica
 Crista intertrochanterica
 Corpus ossis femoris
  Linea aspera
   Labium laterale
   Labium mediale
  Linea pectinea
  Tuberositas glutea
  Facies poplitea
  Linea supracondylaris medialis
  Linea supracondylaris lateralis
 Condylus medialis
  Epicondylus medialis
  Tuberculum adductorium
 Condylus lateralis
  Epicondylus lateralis
 Facies patellaris
 Fossa intercondylaris
 Linea intercondylaris

PATELLA
 Basis patellae
 Apex patellae
 Facies articularis
 Facies anterior

TIBIA
 Condylus medialis
 Condylus lateralis
  Facies articularis fibularis
 Facies articularis superior
 Area intercondylaris anterior
 Area intercondylaris posterior
 Eminentia intercondylaris
  Tuberculum intercondylare mediale
  Tuberculum intercondylare laterale
 Corpus tibiae
  Tuberositas tibiae

Facies medialis
Facies posterior
 Linea musculi solei
Facies lateralis
Margo anterior
Margo medialis
Margo interosseus
Malleolus medialis
 Sulcus malleolaris
 Facies articularis malleoli
Incisura fibularis
Facies articularis inferior

· FIBULA
Caput fibulae
 Facies articularis capitis fibulae
 Apex capitis fibulae
Collum fibulae
Corpus fibulae
 Facies lateralis
 Facies medialis
 Facies posterior
  Crista medialis
 Margo anterior
 Margo interosseus
 Margo posterior
 Malleolus lateralis
  Facies articularis malleoli
  Fossa malleoli lateralis
  Sulcus malleolaris

OSSA TARSI [TARSALIA]

Talus
Caput tali
 Facies articularis navicularis
Collum tali
Corpus tali
Trochlea tali
 Facies superior
 Facies malleolaris medialis
 Facies malleolaris lateralis
  Processus lateralis tali
Facies articularis calcanea posterior
 Sulcus tali
Facies articularis calcanea media
Facies articularis calcanea anterior
Processus posterior tali
 Sulcus tendinis musculi flexoris
  hallucis longi

Tuberculum mediale
Tuberculum laterale
(Os trigonum)

Calcaneus
Tuber calcanei
 Processus medialis tuberis calcanei
 Processus lateralis tuberis calcanei
Tuberculum calcanei
Sustentaculum tali
 Sulcus tendinis musculi flexoris
  hallucis longi
Sulcus calcanei
Sinus tarsi
Facies articularis talaris anterior
Facies articularis talaris media
Facies articularis talaris posterior
Sulcus tendinis musculi peronei
 [fibularis] longi
Trochlea peronealis [fibularis]
Facies articularis cuboidea

Os naviculare
Tuberositas ossis navicularis

Os cuneiforme mediale

Os cuneiforme intermedium

Os cuneiforme laterale

Os cuboideum
 Sulcus tendinis musculi peronei longi
 Tuberositas ossis cuboidei
 Processus calcaneus[24]

OSSA METATARSI [METATARSALIA] (I–V)
 Basis metatarsalis
 Corpus metatarsale
 Caput metatarsale
Tuberositas ossis metatarsalis (primi [I])
Tuberositas ossis metatarsalis (quinti [V])

OSSA DIGITORUM [PHALANGES]
 Phalanx proximalis
 Phalanx media
 Phalanx distalis
  Tuberositas phalangis distalis
 Basis phalangis
 Corpus phalangis

[24] This term denotes the process which projects posteriorly from the plantar or inferior surface of the *Os cuboideum* and supports the anterior extremity of the *Calcaneus*.

Phalanx media
Phalanx distalis
  Tuberositas phalangis distalis
Basis phalangis
Corpus phalangis
Caput phalangis
Ossa sesamoidea

ARTHROLOGIA

ARTICULATIONES FIBROSAE

SYNDESMOSIS
Ligamentum pterygospinale
Ligamentum stylohyoideum
Ligamenta interspinalia
Ligamenta flava
Ligamenta intertransversaria
Ligamenta supraspinalia
Ligamentum nuchae
Syndesmosis [Articulatio] radio-ulnaris
  Membrana interossea antebrachii
  Chorda obliqua
Syndesmosis [Articulatio] tibiofibularis
  Membrana interossea cruris
  Ligamentum tibiofibulare anterius
  Libamentum tibiofibulare posterius

SUTURA
Sutura serrata
Sutura squamosa
Sutura plana
Schindylesis
Suturae cranii [craniales]
  Sutura coronalis
  Sutura sagittalis
  Sutura lambdoidea
  Sutura occipitomastoidea
  Sutura sphenofrontalis
  Sutura spheno-ethmoidalis
  Sutura sphenosquamosa
  Sutura sphenoparietalis
  Sutura squamosa
  (Sutura frontalis [Sutura metopica])
  Sutura parietomastoidea
  (Sutura squamosomastoidea)

Sutura frontonasalis
Sutura fronto-ethmoidalis
Sutura frontomaxillaris
Sutura frontolacrimalis
Sutura frontozygomatica
Sutura zygomaticomaxillaris
Sutura ethmoidomaxillaris
Sutura ethmoidolacrimalis
Sutura sphenovomeriana
Sutura sphenozygomatica
Sutura sphenomaxillaris
Sutura temporozygomatica
Sutura internasalis
Sutura nasomaxillaris
Sutura lacrimomaxillaris
Sutura lacrimoconchalis
Sutura intermaxillaris
Sutura palatomaxillaris
Sutura palato-ethmoidalis
Sutura palatina mediana
Sutura palatina transversa

GOMPHOSIS (ARTICULATIO
  DENTOALVEOLARIS)
Periodontium
  Periodontium protectoris (Gingiva)
  Periodontium insertionis
    Desmodontium
  Cementum
  Os alveolare

ARTICULATIONES CARTILAGINEAE

SYNCHONDROSIS
Synchondroses cranii [craniales]
  Synchondrosis spheno-occipitalis
  Synchondrosis sphenopetrosa
  Synchondrosis petro-occipitalis
  (Synchondrosis intra-occipitalis posterior)
  (Synchondrosis intra-occipitalis anterior)
  Synchondrosis intra-occipitalis
  Synchondrosis spheno-ethmoidalis
Synchondroses sternales
  Synchondrosis xiphisternalis
  (Synchondrosis manubriosternalis)[25]

SYMPHYSIS
Symphysis manubriosternalis[25]
Symphysis intervertebralis

[25] This articulation is at first a synchondrosis and later a symphysis.

Disci intervertebrales
  Annulus [Anulus] fibrosus
  Nucleus pulposus
  Ligamentum longitudinale anterius
  Ligamentum longitudinale posterius
Symphysis pubica
  Ligamentum pubicum superius
  Ligamentum arcuatum pubis
  Discus interpubicus

ARTICULATIONES SYNOVIALES

Articulatio simplex
Articulatio composita
Articulatio plana
Articulatio spheroidea [cotylica]
Articulatio ellipsoidea [condylaris]
Ginglymus
Articulatio bicondylaris
Articulatio trochoidea
Articulatio sellaris[26]
Cartilago articularis
Cavitas articularis
Discus articularis
Meniscus articularis
Labrum articulare
Capsula articularis
  Membrana fibrosa [Stratum fibrosum]
  Membrana synovialis [Stratum synoviale]
    Plica synovialis
      Villi synoviales
  Synovia
Ligamenta
  Ligamenta extracapsularia
  Ligamenta capsularia
  Ligamenta intracapsularia

ARTICULATIONES SYNOVIALES CRANII

ARTICULATIO TEMPOROMANDIBULARIS
  Discus articularis
  Ligamentum laterale
  Ligamentum mediale
  Membrana synovialis superior
  Membrane synovialis inferior

  Ligamentum sphenomandibulare[27]
  Ligamentum stylomandibulare[27]

ARTICULATIO ATLANTO-OCCIPITALIS
  Membrana atlanto-occipitalis anterior
    (Ligamentum atlanto-occipitale
    anterius)
  Membrana atlanto-occipitalis posterior
  Ligamentum atlanto-occipitale laterale

ARTICULATIO ATLANTO-AXIALIS MEDIANA
  Ligamenta alaria
  Ligamentum apicis dentis
  Ligamentum cruciforme atlantis
    Fasciculi longitudinales
  Ligamentum transversum atlantis
  Membrana tectoria

ARTICULATIONES VERTEBRALES[28]

ARTICULATIONES ZYGAPOPHYSIALES

ARTICULATIO LUMBOSACRALIS
  Ligamentum iliolumbale

ARTICULATIO SACROCOCCYGEA
  Ligamentum sacrococcygeum posterius
    [dorsale] superficiale
  Ligamentum sacrococcygeum posterius
    [dorsale] profundum
  Ligamentum sacrococcygeum anterius
    [ventrale]
  Ligamentum sacrococcygeum laterale

ARTICULATIONES THORACIS

ARTICULATIONES COSTOVERTEBRALES

ARTICULATIO CAPITIS COSTAE
  Ligamentum capitis costae radiatum
  Ligamentum capitis costae intra-
    articulare

ARTICULATIO COSTOTRANSVERSARIA
  Ligamentum costotransversarium
  Ligamentum costotransversarium
    superius

---

[26] Since there is good evidence that all articular surfaces are basically sellar or ovoid in their curvatures, the latter term has been tentatively included. Plane surfaces are in fact almost always sellar or ovoid.
[27] Included here only for convenience.
[28] The intervertebral *symphyses* or "discs" are under "Symphysis," page A 22.

Ligamentum costotransversarium laterale
Ligamentum lumbocostale
Foramen costotransversarium

ARTICULATIONES STERNOCOSTALES[29]
Ligamentum sternocostale intra-articulare
Ligamenta sternocostalia radiata
Membrana sterni
Ligamenta costoxiphoidea
Membrana intercostalis externa
Membrana intercostalis interna

ARTICULATIONES COSTOCHONDRALES

ARTICULATIONES INTERCHONDRALES

ARTICULATIONES CINGULI MEMBRI SUPERIORIS[30]

Ligamentum coracoacromiale
Ligamentum transversum scapulae superius
(Ligamentum transversum scapulae
inferius)

ARTICULATIO ACROMIOCLAVICULARIS
Ligamentum acromioclaviculare
Discus articularis
Ligamentum coracoclaviculare
Ligamentum trapezoideum
Ligamentum conoideum

ARTICULATIO STERNOCLAVICULARIS
Discus articularis
Ligamentum sternoclaviculare anterius
Ligamentum sternoclaviculare posterius
Ligamentum costoclaviculare
Ligamentum interclaviculare

ARTICULATIONES MEMBRI SUPERIORIS LIBERI[30]

ARTICULATIO (CAPITIS) HUMERI
Labrum glenoidale
Ligamenta glenohumeralia
Ligamentum coracohumerale

ARTICULATIO CUBITI
Articulatio humero-ulnaris
Articulatio humeroradialis
Articulatio radio-ulnaris proximalis
Ligamentum collaterale ulnare
Ligamentum collaterale radiale
Ligamentum annulare [anulare] radii
Ligamentum quadratum
Membrana interossea antebrachii
Chorda obliqua

ARTICULATIO RADIO-ULNARIS DISTALIS
Discus articularis
Recessus sacciformis
Articulatio radiocarpalis

ARTICULATIONES CARPI
Articulationes intercarpales
Articulatio mediocarpalis
Ligamentum radiocarpale dorsale
Ligamentum radiocarpale palmare
Ligamentum ulnocarpale palmare
Ligamentum carpi radiatum
Ligamentum collaterale carpi ulnare
Ligamentum collaterale carpi radiale
Ligamenta intercarpalia dorsalia
Ligamenta intercarpalia palmaria
Ligamenta intercarpalia interossea
Articulatio ossis pisiformis
Ligamentum pisohamatum
Ligamentum pisometacarpeum
Canalis carpi [carpalis]

ARTICULATIONES CARPOMETACARPALES
Ligamenta carpometacarpalia dorsalia
Ligamenta carpometacarpalia palmaria[31]
Articulatio carpometacarpalis pollicis

ARTICULATIONES INTERMETACARPALES
Ligamenta metacarpalia dorsalia
Ligamenta metacarpalia palmaria[31]
Ligamenta metacarpalia interossea
Spatia interossea metacarpi

ARTICULATIONES METACARPOPHALANGEALES
Ligamenta collateralia

---

[29] The sternal synchondroses are on page A22.
[30] These major divisions of the limbs are introduced merely to assist the reader; some members of the Committee considered them cumbrous and unnecessary. (*See* also page A25).
[31] *Palmaria* was preferred by the Committee, although *ventralia* would accord better with the preceding term.

Ligamenta palmaria
Ligamentum metacarpeum transversum
  profundum

ARTICULATIONES INTERPHALANGEALES
  MANUS
Ligamenta collateralia
Ligamenta palmaria

ARTICULATIONES CINGULI MEMBRI
  INFERIORIS[32]

(Symphysis pubica *see* Page A23)
Membrana obturatoria
Canalis obturatorius
Ligamentum sacrotuberale
  Processus falciformis
Ligamentum sacrospinale
Foramen ischiadicum [sciaticum] majus
Foramen ischiadicum [sciaticum] minus

ARTICULATIO SACROILIACA
Ligamenta sacroiliaca anteriora [ventralia]
Ligamenta sacroiliaca interossea
Ligamenta sacroiliaca posteriora [dorsalia]

ARTICULATIONES MEMBRI
  INFERIORIS LIBERI[32]

ARTICULATIO COXAE
Zona orbicularis
Ligamentum iliofemorale
Ligamentum ischiofemorale
Ligamentum pubofemorale
Labrum acetabulare
  Ligamentum transversum acetabuli
Ligamentum capitis femoris

ARTICULATIO GENUS
Meniscus lateralis
  Ligamentum meniscofemorale anterius
  Ligamentum meniscofemoral posterius
Meniscus medialis
Ligamentum transversum genus
Ligamenta cruciata genus

Ligamentum cruciatum anterius
Ligamentum cruciatum posterius
Plica synovialis infrapatellaris
  Plicae alares
Ligamentum collaterale fibulare
Ligamentum collaterale tibiale
Ligamentum popliteum obliquum
Ligamentum popliteum arcuatum
Ligamentum patellae
Retinaculum patellae mediale
Retinaculum patellae laterale
Corpus adiposum infrapatellare

ARTICULATIO TIBIOFIBULARIS
Ligamentum capitis fibulae anterius
Ligamentum capitis fibulae posterius
Membrana interossea cruris
(Syndesmosis tibiofibularis *see* page A22)

ARTICULATIO TALOCRURALIS
Ligamentum mediale [deltoideum]
  Pars tibionavicularis
  Pars tibiocalcaneus
  Pars tibiotalaris anterior
  Pars tibiotalaris posterior
Ligamentum talofibulare anterius
Ligamentum talofibulare posterius
Ligamentum calcaneofibulare

ARTICULATIONES PEDIS

ARTICULATIO TARSI TRANSVERSA

ARTICULATIO TALOCALCANEONAVICULARIS
Articulatio subtalaris[33]
Ligamentum talocalcaneare laterale
Ligamentum talocaleaneare mediale

ARTICULATIO CALCANEOCUBOIDEA

ARTICULATIO CUNEONAVICULARIS
Ligamenta tarsi interossea
  Ligamentum talocalcaneare interosseum
  Ligamentum cuneocuboideum interosseum
  Ligamenta intercuneiformia interossea
Ligamenta tarsi dorsalia
  Ligamentum talonaviculare

---

[32] *See* footnote 30.
[33] This term was preferred in the third edition because of its prevalence in clinical usage, although *A. talocalcanea* (B.N.A.) is more regular and informative. Perhaps the latter term should be reinstated as an official synonym.

Ligamenta intercuneiformia dorsalia
Ligamentum cuneocuboideum dorsale
Ligamentum cuboideonaviculare dorsale
Ligamentum bifurcatum
    Ligamentum calcaneonaviculare
    Ligamentum calcaneocuboideum
Ligamenta cuneonavicularia dorsalia
Ligamenta tarsi plantaria
Ligamentum plantare longum
Ligamentum calcaneocuboideum plantare
Ligamentum calcaneonaviculare plantare
Ligamenta cuneonavicularia plantaria
Ligamentum cuboideonaviculare plantare
Articulationes intercuneiformes
Ligamenta intercuneiformia plantaria
Articulatio cuneocuboidea
Ligamentum cuneocuboideum plantare
Articulationes intercuneiformes
Articulatio cuneo cuboidea
Ligamenta intercuneiformia plantaria
Ligamentum cuneocuboideum plantare

ARTICULATIONES TARSOMETATARSALES
Ligamenta tarsometatarsalia dorsalia
Ligamenta tarsometatarsalia plantaria
Ligamenta cuneometatarsalia interossea

ARTICULATIONES INTERMETATARSALES
Ligamenta metatarsalia interossea
Ligamenta metatarsalia dorsalia
Ligamenta metatarsalia plantaria
Spatia interossea metatarsi

ARTICULATIONES METATARSOPHALANGEALES
Ligamenta collateralia
Ligamenta plantaria
Ligamentum metatarsale transversum
  profundum

ARTICULATIONES INTERPHALANGEALES PEDIS
Ligamenta collateralia

Ligamenta plantaria

MYOLOGIA

Musculus[34]
  Caput
  Venter
Musculus fusiformis
Musculus quadratus
Musculus triangularis
Musculus unipennatus[35]
Musculus bipennatus[35]
Musculus multipennatus
Musculus sphincter
Musculus orbicularis
Musculus cruciatus
Musculus cutaneus
Tendo
Vagina tendinis
  Stratum fibrosum
  Stratum synoviale
    Vagina synovialis tendinis
  Mesotendineum
  Peritendineum
Aponeurosis
Epimysium
Perimysium
Endomysium
Fascia
  Superficialis
  Profunda
Intersectio tendinea
Arcus tendineus
Trochlea muscularis
Bursa synovialis

MUSCULI CAPITIS[36, 37]

---

[34] Attempts to find terms for the 'origins and insertions' of muscles were unsuccessful. Widespread disagreement in the I.A.N.C. and its Subcommittees made it impossible to reach decisions. Suggestions made included *Insertio proximale* [*origo*], *Insertio distale* [*terminatio*], *Punctum fixum*, *Punctum mobile*, etc.

[35] Logically these terms should be *semipennatus* and *pennatus*, if they are to accord with the usual structure of a feather (*penna*).

[36] It is not possible to arrange the muscles of the head and neck in a completely satisfactory manner. Many of the muscles here can be considered to belong to both regions. Further difficulties arise in classifying muscles as "facial" or masticatory. It is essential to appreciate that the arrangement adopted, as in many other places in this volume, is purely to aid reference.

[37] Many anatomists suggested the transference of the extraocular, lingual, palatal, pharyngeal and laryngeal muscles from *Organa Sensuum* and *Splanchnologia*. Others disagreed with this suggestion. The Committee therefore decided to include these muscles in *Myologia* and in their original sections.

MUSCULI BULBI[37] (*see* page A81)

MUSCULI OSSICULORUM
AUDITORIORUM (*see* page A84)

MUSCULI SUBOCCIPITALES

M. rectus capitis anterior
M. rectus capitis posterior major
M. rectus capitis posterior minor
M. rectus capitis lateralis
M. obliquus capitis superior
M. obliquus capitis inferior

MUSCULI FACIALES ET
  MASTICATORES

M. epicranius
  M. occipitofrontalis
    Venter frontalis
    Venter occipitalis
  M. temporoparietalis
  Galea aponeurotica [Aponeurosis
    epicranialis]
M. procerus
M. nasalis
  Partes transversa et alaris
M. depressor septi
M. orbicularis oculi
  Pars palpebralis
  Pars orbitalis
  Pars lacrimalis
M. corrugator supercilii
M. depressor supercilii
M. auricularis anterior
M. auricularis superior
M. auricularis posterior
M. orbicularis oris
  Pars marginalis
  Pars labialis
M. depressor anguli oris
M. transversus menti
M. risorius
M. zygomaticus major
M. zygomaticus minor
M. levator labii superioris
M. levator labii superioris alaeque nasi
M. depressor labii inferioris
M. levator anguli oris
M. buccinator
M. mentalis

M. masseter
  Pars superficialis
  Pars profunda
M. temporalis
M. pterygoideus lateralis
M. pterygoideus medialis
Fascia buccopharyngea
Fascia masseterica
Fascia parotidea
Fascia temporalis
  Lamina superficialis
  Lamina profunda

MUSCULI LINGUAE (*see* page A34)

MUSCULI PALATI ET FAUCIUM
  (*see* page A34)

MUSCULI COLLI

Platysma
M. longus colli
M. longus capitis
M. scalenus anterior
M. scalenus medius
M. scalenus posterior
(M. scalenus minimus)
M. sternocleidomastoideus

MUSCULI SUPRAHYOIDEI

M. digastricus
  Venter anterior
  Venter posterior
M. stylohyoideus
M. mylohyoideus
M. geniohyoideus

MUSCULI INFRAHYOIDEI

M. sternohyoideus
M. omohyoideus
  Venter superior
  Venter inferior
M. sternothyroideus
M. thyrohyoideus
(M. levator glandulae thyroideae)

FASCIA CERVICALIS

  Lamina superficialis

Lamina pretrachealis
Lamina prevertebralis
Vagina carotica

TUNICA MUSCULARIS
PHARYNGIS
(*see* page A34)

MUSCULI LARYNGIS
(*see* page A39)

MUSCULI DORSI

Fascia nuchae [nuchalis]
M. trapezius
(M. transversus nuchae)
M. latissimus dorsi
M. rhomboideus major
M. rhomboideus minor
M. levator scapulae
M. serratus posterior inferior
M. serratus posterior superior
M. splenius capitis
M. splenius cervicis

MUSCULUS ERECTOR SPINAE

Fascia thoracolumbalis
M. iliocostalis
　M. iliocostalis lumborum
　M. iliocostalis thoracis
　M. iliocostalis cervicis
M. longissimus
　M. longissimus thoracis
　M. longissimus cervicis
　M. longissimus capitis
M. spinalis
　M. spinalis thoracis
　M. spinalis cervicis
　M. spinalis capitis

MUSCULI TRANSVERSOSPINALES

M. semispinalis
　M. semispinalis thoracis
　M. semispinalis cervicis

M. semispinalis capitis
Mm. multifidi
Mm. rotatores
　Mm. rotatores cervicis
　Mm. rotatores thoracis
　Mm. rotatores lumborum

MUSCULI INTERSPINALES

Mm. interspinales cervicis
Mm. interspinalis thoracis
Mm. interspinales lumborum

MUSCULI INTERTRANSVERSARII

Mm. intertransversarii laterales lumborum
Mm. intertransversarii mediales lumborum
Mm. intertransversarii thoracis
Mm. intertransversarii posteriores cervicis
　Pars medialis
　Pars lateralis
Mm. intertransversarii anteriores cervicis

MUSCULI THORACIS

(M. sternalis)
M. pectoralis major
　Pars clavicularis
　Pars sternocostalis
　Pars abdominalis
M. pectoralis minor
M. subclavius
Fascia pectoralis
Fascia clavipectoralis
M. serratus anterior
Mm. levatores costarum
　Mm. levatores costarum longi
　Mm. levatores costarum breves
Mm. intercostales externi
Membrana intercostalis externa
Mm. intercostales interni
Membrana intercostalis interna
Mm. intercostales intimi[38]
Mm. subcostales
M. transversus thoracis
Fascia thoracica[39]
Fascia endothoracica[39]

---

[38] Separated from the *Mm. intercostales interni* by the intercostal nerves.
[39] Suggestions made by the Nomenclatural Commission of the U.S.S.R. and accepted by the Committee.

DIAPHRAGMA

Pars lumbalis
  Crus dextrum
  Crus sinistrum
Pars costalis
Pars sternalis
Hiatus aorticus
Hiatus oesophageus [esophageus]
Centrum tendineum
Foramen venae cavae
Ligamentum arcuatum mediale[40]
Ligamentum arcutum laterale[40]
Ligamentum arcuatum medianum[40]

MUSCULI ABDOMINIS

M. rectus abdominis
  Intersectiones tendineae
  Vagina musculi recti abdominis
    Lamina anterior
    Lamina posterior
      Linea arcuata
M. pyramidalis
M. obliquus externus abdominis
  Ligamentum inguinale [Arcus inguinalis]
    Ligamentum lacunare
      Ligamentum pectineale
    Ligamentum reflexum
Annulus [Anulus] inguinalis superficialis
  Crus mediale
  Crus laterale
  Fibrae intercrurales
M. obliquus internus abdominis
  M. cremaster
M. transversus abdominis
  Falx inguinalis [Tendo conjunctivus]
  Annulus [Anulus] inguinalis profundus
Linea alba
  Annulus [Anulus] umbilicalis
  Adminiculum lineae albae
Ligamentum suspensorium penis/clitoridis
Ligamentum fundiforme penis
Trigonum lumbale
Fascia transversalis
  Annulus [Anulus] inguinalis profundus
Canalis inguinalis
Ligamentum interfoveolare
M. quadratus lumborum

MUSCULI DIAPHRAGMATIS PELVIS

M. levator ani (*see* page A45)
  M. pubococcygeus
  M. levator prostatae
  M. pubovaginalis
  M. puborectalis
  M. iliococcygeus
(Arcus tendineus musculi levatoris ani)
M. coccygeus
(M. sacrococcygeus ventralis)
(M. sacrococcygeus dorsalis)
M. sphincter ani externus (*see* pages A36, 45)
  Pars subcutanea
  Pars superficialis
  Pars profunda
Ligamentum anococcygeum
Fascia pelvis
  Fascia pelvis parietalis
    Fascia obturatoria
  Fascia pelvis visceralis
    Fascia prostatae
      Septum rectovesicale
      Septum rectovaginale
Fascia diaphragmatis pelvis superior
Arcus tendineus fasciae pelvicae
Ligamentum puboprostaticum
      [pubovesicale]
Fascia diaphragmatis pelvis inferior

MUSCULI DIAPHRAGMATIS
UROGENITALIS (*see* page A45)

MUSCULI MEMBRI SUPERIORIS

M. deltoideus
M. supraspinatus
M. infraspinatus
M. teres minor
M. teres major
M. subscapularis
M. biceps brachii
  Caput longum
    Vagina tendinis intertubercularis
  Caput breve
  Aponeurosis musculi bicipitis brachii
    [Apon. bicipitalis]

---

[40] As some anatomists have indicated, these are not strictly ligaments, but no suitable alternative term has been suggested, except *Arcus musculi psoatis*, etc.

M. coracobrachialis
M. brachialis
M. triceps brachii
  Caput longum
  Caput laterale
  Caput mediale
M. anconeus
M. articularis cubiti
M. pronator teres
  Caput humerale
  Caput ulnare
M. flexor carpi radialis
M. palmaris longus
M. flexor carpi ulnaris
  Caput humerale
  Caput ulnare
M. flexor digitorum superficialis
  Caput humero-ulnare
  Caput radiale
M. flexor digitorum profundus
M. flexor pollicis longus
M. pronator quadratus
M. brachioradialis
M. extensor carpi radialis longus
M. extensor carpi radialis brevis
M. extensor digitorum
  Connexus [Conexus] intertendineus
M. extensor digiti minimi
M. extensor carpi ulnaris
  Caput humerale
  Caput ulnare
M. supinator
M. abductor pollicis longus
M. extensor pollicis brevis
M. extensor pollicis longus
M. extensor indicis
M. palmaris brevis
M. abductor pollicis brevis
M. flexor pollicis brevis
  Caput superficiale
  Caput profundum
M. opponens pollicis
M. adductor pollicis
  Caput obliquum
  Caput transversum
M. abductor digiti minimi
M. flexor digiti minimi brevis
M. opponens digiti minimi
Mm. lumbricales
Mm. interossei dorsales
Mm. interossei palmares
Fascia axillaris

Fascia deltoidea
Fascia brachii (brachialis)
Septum intermusculare brachii mediale
Septum intermusculare brachii laterale
Fascia antebrachii
Fascia dorsalis manus
Retinaculum extensorum
Ligamentum metacarpale transversum
  superficiale
Aponeurosis palmaris
  Fasciculi transversi
Retinaculum flexorum
Canalis carpi (*see* page A24)
Chiasma tendinum

MUSCULI MEMBRI INFERIORIS

M. iliopsoas
  M. iliacus
  M. psoas major
(M. psoas minor)
M. gluteus maximus
M. gluteus medius
M. gluteus minimus
M. tensor fasciae latae
M. piriformis
M. obturator internus
M. gemellus superior
M. gemellus inferior
M. quadratus femoris
M. sartorius
M. quadriceps femoris
  M. rectus femoris
    Caput rectum
    Caput reflexum
  M. vastus lateralis
  M. vastus intermedius
  M. vastus medialis
M. articularis genus
M. pectineus
M. adductor longus
M. adductor brevis
M. adductor magnus
M. gracilis
M. obturator externus
M. biceps femoris
  Caput longum
  Caput breve
M. semitendinosus
M. semimembranosus
M. tibialis anterior

M. extensor digitorum longus
M. peroneus tertius [M. fibularis tertius]
M. extensor hallucis longus
M. peroneus longus [M. fibularis longus]
M. peroneus brevis [M. fibularis brevis]
M. triceps surae
  M. gastrocnemius
    Caput laterale
    Caput mediale
  M. soleus
    Arcus tendineus musculi solei
  Tendo calcaneus (Achilles)
M. plantaris
M. popliteus
M. tibialis posterior
M. flexor digitorum longus
M. flexor hallucis longus
M. extensor hallucis brevis
M. extensor digitorum brevis
M. abductor hallucis
  Caput mediale
  Caput laterale
M. flexor hallucis brevis
M. adductor hallucis
  Caput obliquum
  Caput transversum
M. abductor digiti minimi
(M. opponens digiti minimi)
M. flexor digiti minimi brevis
M. flexor digitorum brevis
M. quadratus plantae [M. flexor
    accessorius]
Mm. lumbricales
Mm. interossei dorsales
Mm. interossei plantares
Fascia lata
  Tractus iliotibialis
Septum intermusculare femoris laterale
Septum intermusculare femoris mediale
Canalis adductorius
Hiatus tendineus [adductorius]
Fascia iliaca
Lacuna musculorum
Arcus iliopectineus
Lacuna vasorum
Trigonum femorale
Canalis femoralis
Annulus [Anulus] femoralis

Septum femorale
Hiatus saphenus
  Margo falciformis
    Cornu superius
    Cornu inferius
  Fascia cribrosa
Fascia cruris
Septum intermusculare cruris anterius
Septum intermusculare cruris posterius
Retinaculum musculorum extensorum
    superius
Retinaculum musculorum flexorum
Retinaculum musculorum extensorum
    inferius
Retinaculum musculorum peroneorum
    [fibularium] superius
Retinaculum musculorum peroneorum
    [fibularium] inferius
Fascia dorsalis pedis
Aponeurosis plantaris
  Fasciculi transversi
Ligamentum metatarsale transversum
    superficiale

BURSAE ET VAGINAE SYNOVIALES[41]

Bursa subcutanea
Bursa submuscularis
Bursa subfascialis
Bursa subtendinea

Bursa musculi tensoris veli palatini
Bursa subcutanea prominentiae laryngealis
Bursa infrahyoidea
Bursa retrohyoidea
Bursa subtendinea musculi trapezii
(Bursa subcutanea acromialis)
Bursa subacromialis
Bursa subdeltoidea
(Bursa musculi coracobrachialis)
Bursa subtendinea musculi infraspinati
Bursa subtendinea musculi subscapularis
Bursa subtendinea musculi teretis majoris
Bursa subtendinea musculi latissimi dorsi
Bursa subcutanea olecrani
(Bursa intratendinea olecrani)
Bursa subtendinea musculi tricipitis brachii
Bursa bicipitoradialis

[41] Many criticisms on the siting of these structures were received. Some anatomists considered that certain, if not all, *bursae* and *vaginae* should be sited directly with the associated muscles. The Committee decided to take no action for the present.

(Bursa cubitalis interossea)

Vaginae fibrosae digitorum manus
Pars annularis [anularis] vaginae fibrosae
Pars cruciformis vaginae fibrosae
Vaginae synoviales digitorum manus
Vincula tendinum
Vinculum longum
Vinculum breve
Vagina tendinum musculorum abductoris
longi et extensoris brevis pollicis
Vagina tendinum musculorum extensorum
carpi radialium
Vagina tendinis musculi extensoris pollicis
longi
Vagina tendinum musculorum extensoris
digitorum et extensoris indicis
Vagina tendinis musculi extensoris carpi
ulnaris
Bursa musculi extensoris carpi radialis
brevis
Vagina tendinis musculi flexoris
carpi radialis
Vagina communis musculorum
flexorum
Vagina tendinis musculi flexoris
pollicis longi
Vaginae tendinum digitorum
manus

Bursa subcutanea trochanterica
Bursa trochanterica musculi glutei maximi
Bursae trochantericae musculi glutei medii
Bursa trochanterica musculi glutei minimi
Bursa musculi piriformis
Bursa ischiadica [sciatica] musculi
obturatoris interni
Bursa subtendinea musculi obturatoris
interni
Bursae intermusculares musculorum
gluteorum
Bursa ischiadica [sciatica] musculi glutei
maximi
(Bursa iliopectinea)
Bursa subtendinea iliaca
Bursa musculi bicipitis femoris superior
Bursa subcutanea prepatellaris
(Bursa subfascialis prepatellaris)
(Bursa subtendinea prepatellaris)
Bursa suprapatellaris
Bursa subcutanea infrapatellaris

Bursa infrapatellaris profunda
Bursa subcutanea tuberositatis tibiae
Bursae subtendineae musculi sartorii
Bursa anserina
Bursa subtendinea musculi bicipitis femoris
inferior
Recessus subpopliteus
Bursa subtendinea musculi gastrocnemii
lateralis
Bursa subtendinea musculi gastrocnemii
medialis
Bursa musculi semimembranosi
Bursa subcutanea malleoli lateralis
Bursa subcutanea malleoli medialis

Vagina tendinis musculi tibialis anterioris
Vagina tendinis musculi extensoris hallucis
longi
Vagina tendinum musculi extensoris
digitorum pedis longi
Vagina tendinum musculi flexoris
digitorum pedis longi
Vagina tendinis musculi tibialis
posterioris
Vagina tendinis musculi flexoris
hallucis longi
Vagina musculorum peroneorum
[fibularium] communis
Bursa subtendinea musculi tibialis
anterioris
Bursa subcutanea calcanea
Bursa tendinis calcanei
Vagina tendinis musculi peronei
[fibularis] longi plantaris
Vaginae tendinum digitorum pedis
Vaginae synoviales digitorum pedis
Vincula tendinum
Vaginae fibrosae digitorum pedis
Pars annularis [anularis] vaginae fibrosae
Pars cruciformis vaginae fibrosae

SPLANCHNOLOGIA

APPARATUS DIGESTORIUS
[SYSTEMA DIGESTORIUM]

CAVITAS ORIS[42]

---

[42]   The Committee have followed the Histology Subcommittee in substituting *Cavitas* for *Cavum* throughout this section.

VESTIBULUM ORIS
  Rima oris
  Labia oris
  Labium superius
    Philtrum
      Tuberculum
  Labium inferius
  Commissura labiorum
  Angulus oris
  Bucca
  Corpus adiposum buccae

CAVITAS ORIS PROPRIA
Palatum
  Palatum durum[43]
  Palatum molle [Velum palatinum]
  Raphe palati

TUNICA MUCOSA ORIS
Frenulum labii superioris
Frenulum labii inferioris
Gingivae
  Margo gingivalis
  Papilla gingivalis [interdentalis]
  Sulcus gingivalis
Caruncula sublingualis
Plica sublingualis
Papilla parotidea
Plicae palatinae transversae
Papilla incisiva

GLANDULAE ORIS

GLANDULAE SALIVARIAE MAJORES

Glandula parotidea
  Pars superficialis
  Pars profunda
  Glandula parotidea accessoria
  Ductus parotideus

Glandula sublingualis
  Ductus sublingualis major
  Ductus sublinguales minores

Glandula submandibularis
  Ductus submandibularis

GLANDULAE SALIVARIAE MINORES[44]
  Gll. labiales

Gll. buccales
Gll. molares
Gll. palatinae
Gll. linguales
  Gll. lingualis anterior

DENTES

Corona dentis[45]
  Cuspis dentis
    Apex cuspidis
  Tuberculum dentis
  Crista transversalis
  Crista triangularis
Corona clinica
Cervix dentis
Radix dentis
  Apex radicis dentis
Radix clinica
Facies occlusalis
Facies vestibularis [facialis]
Facies lingualis
Facies contactus
  Facies mesialis
  Facies distalis
Cingulum
Crista marginalis
Margo incisalis
Cavitas dentis [pulparis]
  Cavitas coronae
  Canalis radicis dentis
    Foramen apicis dentis
Pulpa dentis
  Pulpa coronalis
  Pulpa radicularis
Papilla dentis
Dentinum
Enamelum
Cementum
Periodontium
Arcus dentalis superior
Arcus dentalis inferior
Dentes incisivi
Dentes canini
Dentes premolares
Dentes molares
  Dens serotinus (molaris tertius)
Dentes decidui
Dentes permanentes

---

[43] This term must not be confused with *Palatum osseum* (see page A 12).
[44] Terms for *Glandulae salivariae* have been rearranged to accord with *Nomina Histologica*.
[45] Several histological terms (e.g. *canaliculi dentales*) have been removed.

NOMINA ANATOMICA

Diastema

LINGUA

Corpus linguae
Radix linguae
Dorsum linguae
  Pars presulcalis [anterior]
  Pars postsulcalis [posterior]
Facies inferior linguae
  Plica fimbriata
Margo linguae
Apex linguae
Tunica mucosa linguae
Frenulum linguae
Papillae linguales
  Papillae filiformes
  Papillae conicae
  Papillae fungiformes
  Papillae vallatae
  Papillae lentiformes
  Papillae foliatae
Sulcus medianus linguae
Sulcus terminalis
Foramen caecum [cecum] linguae
  Ductus thyroglossalis
Tonsilla lingualis
Septum linguae
Aponeurosis linguae
Folliculi linguales

MUSCULI LINGUAE
M. genioglossus
M. hyoglossus
M. chondroglossus
M. styloglossus
M. longitudinalis superior
M. longitudinalis inferior
M. transversus linguae
M. verticalis linguae

FAUCES

Isthmus faucium
Palatum molle [Velum palatinum]
  Uvula palatina
  Arcus palatoglossus
  Arcus palatopharyngeus
Plica salpingopalatina
Tonsilla palatina

Fossulae tonsillares
  Cryptae tonsillares
Capsula tonsillaris
Plica semilunaris
Plica triangularis
Fossa tonsillaris
Fossa supratonsillaris

MUSCULI PALATI ET FAUCIUM
Aponeurosis palatina
M. levator veli palatini
M. tensor veli palatini
M. uvulae
M. palatoglossus
M. palatopharyngeus

PHARYNX

CAVITAS PHARYNGIS
Pars nasalis pharyngis
Fornix pharyngis
Tonsilla pharyngealis [adenoidea][46]
  Fossulae tonsillares
    Cryptae tonsillares
(Bursa pharyngealis)
Ostium pharyngeum tubae auditoniae
  Torus tubarius
    Plica salpingopharyngea
    Plica salpingopalatina
  Torus levatorius
  Tonsilla tubaria
  Recessus pharyngeus [-gialis]
Pars oralis pharyngis
Vallecula epiglottica
  Plica glossoepiglottica mediana
  Plica glossoepiglottica lateralis
Pars laryngea pharyngis
Recessus piriformis
  Plica nervi laryngei
Fascia pharyngobasilaris
Tela submucosa
Tunica mucosa
  Glandulae pharyngeae [-geales]

TUNICA MUSCULARIS PHARYNGIS
Raphe pharyngis
Raphe pterygomandibularis
M. constrictor pharyngis superior
  Pars pterygopharyngea

[46]   *T. pharyngea* has been changed to *T. pharyngealis* [*adenoidea*] to accord with *Nomina Histologica.*

A 34

Pars buccopharyngea
Pars mylopharyngea
Pars glossopharyngea
M. constrictor pharyngis medius
Pars chondropharyngea
Pars ceratopharyngea
M. constrictor pharyngis inferior
Pars thyropharyngea
Pars cricopharyngea
M. stylopharyngeus
M. salpingopharyngeus
Fascia buccopharyngealis
Spatium peripharyngeum
   Spatium retropharyngeum
   Spatium lateropharyngeum

OESOPHAGUS [ESOPHAGUS]

Pars cervicalis
Pars thoracica
Pars abdominalis
Tunica adventitia
Tunica muscularis
   Tendo cricoesophageus
M. bronchoesophageus
M. pleuroesophageus
Tela submucosa
Tunica mucosa
   Lamina muscularis mucosae
Gll. oesophageae [esophageae]

GASTER [VENTRICULUS][47]

Paries anterior
Paries posterior
Curvatura gastrica [ventriculi] major
Curvatura gastrica [ventriculi] minor
   Incisura angularis
Pars cardiaca
   Ostium cardiacum
Fundus gastricus [ventricularis]
Fornix gastricus [ventricularis][48]
   Incisura cardiaca
Corpus gastricum [ventriculare]
   Canalis gastricus [ventricularis]

Pars pylorica
   Antrum pyloricum
   Canalis pyloricus
Pylorus
   Ostium pyloricum
Tunica serosa
Tela subserosa
Tunica muscularis
   Stratum longitudinale
   Stratum circulare
      M. sphincter pyloricus
   Fibrae obliquae
Tela submucosa
Tunica mucosa
   Plicae gastricae
   Lamina muscularis mucosae
   Areae gastricae
   Plicae villosae
   Foveolae gastricae

INTESTINUM TENUE

Tunica serosa
Tela subserosa
Tunica muscularis
   Stratum longitudinale
   Stratum circulare
Tela submucosa
Tunica mucosa
   Lamina muscularis mucosae
   Plicae circulares
   Villi intestinales
   Gll. intestinales
   Folliculi lymphatici solitarii
   Folliculi lymphatici aggregati

DUODENUM
Pars superior
   Ampulla[49]
Pars descendens
Pars horizontalis [inferior]
Pars ascendens
Flexura duodeni superior
Flexura duodeni inferior
Flexura duodenojejunalis
M. suspensorius duodeni

---

[47] Since almost everything associated with the "ventriculus" is *gastric*, the Committee preferred *gaster*.

[48] The Russian Nomenclatural Commission suggested *fornix ventriculi* as a useful term in radiological anatomy.

[49] A new term to cover the "duodenal cap" of radiological anatomy. A true duodenal ampulla is present in some domesticated mammals.

Plica longitudinalis duodeni
Papilla duodeni major
Papilla duodeni minor
Gll. duodenales

JEJUNUM

ILEUM

INTESTINUM CRASSUM

CAECUM [CECUM][50]
Valva ileocaecalis [-cecalis] [Valva ilealis]
Papilla ileocaecalis [-cecalis] [Papilla ilealis]
Ostium valvae ilealis
Frenulum valvae ilealis
Ostium ileocaecale [-cecale]
Appendix vermiformis
    Ostium appendicis vermiformis
    Folliculi lymphatici aggregati
        appendicis vermiformis

COLON
Colon ascendens
Flexura coli dextra
Colon transversum
Flexura coli sinistra
Colon descendens
Colon sigmoideum
Plicae semilunares coli
Haustra coli
Appendices epiploicae [omentales]
Tunica muscularis
    Stratum longitudinale
        Taeniae [Teniae] coli
        Taenia [Tenia]
           mesocolica
           omentalis
           libera
Stratum circulare

RECTUM[50]
Flexura sacralis
Flexura perinealis
Ampulla recti

Tunica muscularis
    Stratum longitudinale
    Stratum circulare
M. sphincter ani internus
M. rectococcygeus
M. rectourethralis
M. rectovesicalis
Plicae transversales recti

CANALIS ANALIS[51]
Columnae anales
Sinus anales
Valvulae anales
Linea anorectalis
Pecten analis
Linea anocutanea
M. sphincter ani externus
(*see* pages 29, 45)
Anus

HEPAR

Facies diaphragmatica
    Pars superior
        Impressio cardiaca
    Pars anterior
    Pars dextra
    Pars posterior
    Area nuda
        Sulcus venae cavae
        Fissura ligamenti venosi
        Ligamentum venosum
Facies visceralis
    Fossa vesicae biliaris
    Fissura ligamenti teretis
        Lig. teres hepatis
    Porta hepatis
    Tuber omentale
    Impressio oesophageale [eso-]
    Impressio gastrica
    Impressio duodenalis
    Impressio colica
    Impressio renalis
    Impressio suprarenalis
Margo inferior
    Incisura ligamenti teretis

---

[50] These terms are really adjectives qualifying *intestinum*. Custom has made them nouns (as in the case of *duodenum*). The Committee accepted this as *fait accompli*.

[51] The *Zona haemorrhoidalis* of the third edition has been omitted by general agreement. It is a pathological concept. The associated veins and venous plexus are no longer qualified as "haemorrhoidal."

Lobi hepatis[52]
  Lobus hepatis dexter
    Segmentum anterius
    Segmentum posterius
  Lobus hepatis sinister
    Segmentum mediale
      Pars quadrata
    (Appendix fibrosa hepatis)
    Segmentum laterale
  Lobus quadratus
  Lobus caudatus
    Processus papillaris
    Processus caudatus
Tunica serosa
Tela subserosa
Tunica fibrosa
  Capsula fibrosa perivascularis[53]
Lobuli hepatis
Arteriae interlobulares
Venae interlobulares
Venae centrales
Ductuli interlobulares
Ductuli biliferi
Ductus hepaticus communis[54]
  Ductus hepaticus dexter
    Ramus anterior
    Ramus posterior
  Ductus hepaticus sinister
    Ramus lateralis
    Ramus medialis
Ductus lobi caudati dexter
Ductus lobi caudati sinister

VESICA BILIARIS [FELLEA][55]
Fundus vesicae biliaris
Corpus vesicae biliaris
Collum vesicae biliaris

Tunica serosa vesicae biliaris
Tela subserosa vesicae biliaris
Tunica muscularis vesicae biliaris
Tunica mucosa vesicae biliaris
  Plicae tunicae mucosae vesicae biliaris
Ductus cysticus
  Plica spiralis

DUCTUS CHOLEDOCHUS [BILIARIS]
M. sphincter ductus choledochi[56]
Ampulla hepatopancreatica
  M. sphincter ampullae hepatopancreaticae
    [Sphincter ampullae][56]
Gll. mucosae biliosae

PANCREAS

Caput pancreatis
  Processus uncinatus
  Incisura pancreatis
Corpus pancreatis
  Facies anterior
  Facies posterior
  Facies inferior
  Margo superior
  Margo anterior
  Margo inferior
  Tuber omentale
Cauda pancreatis
Ductus pancreaticus
  M. sphincter ductus pancreatici
Ductus pancreaticus accessorius
(Pancreas accessorium)

APPARATUS RESPIRATORIUS
[SYSTEMA RESPIRATORIUM]

---

[52] The division into lobes is based on superficial features; division into segments is based upon the ramification of the bile ducts and hepatic vessels. The two modes do not entirely coincide. Moreover, there is not yet complete agreement between authorities on segmentation. Therefore the Committee decided to leave these terms unchanged, though admitting that they are in some details unsatisfactory. The principal schemes so far propounded are: by: C.-H. Hjortsjö (*Acta Anat. 11*, 599, 1951, and *Lunds Univ. Arsskr.* N. F. *44*, 1948), J. E. Healey and P. C. Schroy (*Arch. Surg. Chicago 66*, 599, 1953), and C. Couinaud (*Presse méd. 62*, 709, 1954).

[53] Glisson's capsule.

[54] More detailed subdivision of the biliary ducts was not approved by the Committee. *See* footnote 52.

[55] A majority of the Commitee recommended the substitution of the internationally familiar *biliaris* for *fella*. The latter is generally unfamiliar, and is not in wide clinical usage.

[56] These sphincters have been described. Incidentally, *sphincter* is a noun, not an adjective. (A reversal of the more frequent change, noun → adjective. See footnote 50.) *Musculus sphincter* is hence a tautological expression.

NASUS EXTERNUS[57]

Radix nasi
Dorsum nasi
Apex nasi
Alae nasi
Cartilagines nasi
  Cartilago nasi lateralis
  Cartilago alaris major
    Crus mediale
    Crus laterale
  Cartilagines alares minores
  Cartilagines nasales accessoriae
  Cartilago septi nasi
    Processus posterior [sphenoidalis]
  Cartilago vomeronasalis
Pars mobilis septi nasi

CAVITAS NASI
Nares
Choanae
Septum nasi
  Pars membranacea
  Pars cartilaginea
  Pars ossea
  Organum vomeronasale
Vestibulum nasi
Limen nasi
Sulcus olfactorius
Concha nasalis superior
Concha nasalis media
Concha nasalis inferior
Regio respiratoria
Regio olfactoria
Gll. nasales
Plexus cavernosi concharum
Agger nasi
Recessus sphenoethmoidalis
Meatus nasi superior
Meatus nasi medius
  Atrium meatus medii
  Bulla ethmoidalis
  Infundibulum ethmoidale
    Hiatus semilunaris
Meatus nasi inferior
Meatus nasopharyngeus
(Ductus incisivus)

SINUS PARANASALES
Sinus maxillaris
Sinus sphenoidalis
Sinus frontalis

Sinus ethmoidales[58]
  anteriores
  medii
  posteriores

LARYNX

CARTILAGINES LARYNGIS

CARTILAGO THYROIDEA
Prominentia laryngea
Lamina dextra/sinistra
Incisura thyroidea superior
Incisura thyroidea inferior
Tuberculum thyroideum superius
Tuberculum thyroideum inferius
Linea obliqua
Cornu superius
Cornu inferius
(Foramen thyroideum)
Membrana thyrohyoidea
  Lig. thyrohyoideum medianum
  Lig. thyrohyoideum laterale
    Cartilago triticea

CARTILAGO CRICOIDEA
  Arcus cartilaginis cricoideae
  Lamina cartilaginis cricoideae
    Facies articularis arytenoidea
    Facies articularis thyroidea
  Articulatio cricothyroidea
    Capsula articularis cricothyroidea
    Lig. ceratocricoideum
  Lig. cricothyroideum medianum
  Lig. cricotracheale

CARTILAGO ARYTENOIDEA
Facies articularis
Basis cartilaginis arytenoideae
Facies anterolateralis
  Processus vocalis
  Crista arcuata
  Colliculus
  Fovea oblonga
  Fovea triangularis
Facies medialis
Facies posterior
  Apex cartilaginis arytenoideae
  Processus muscularis
Articulatio cricoarytenoidea

---

[57] *Nasus externus* now precedes *Cavitas nasi*,—a more logical order.
[58] The subdivisions may be *Sinus* or *Cellulae anteriores*, etc.

Capsula articularis cricoarytenoidea
Lig. cricoarytenoideum posterius
Lig. cricopharyngeum
(Cartilago sesamoidea)

CARTILAGO CORNICULATA
Tuberculum corniculatum

CARTILAGO CUNEIFORMIS
Tuberculum cuneiforme

EPIGLOTTIS
Petiolus epiglottidis
Tuberculum epiglotticum
Cartilago epiglottica
Lig. thyroepiglotticum
Lig. hyoepiglotticum

MUSCULI LARYNGIS
M. aryepiglotticus
M. cricothyroideus
 Pars recta
 Pars obliqua
M. cricoarytenoideus posterior
(M. ceratocricoideus)
M. cricoarytenoideus lateralis
M. vocalis
M. thyroepiglotticus
M. thyroarytenoideus
M. arytenoideus obliquus
M. arytenoideus transversus

CAVITAS LARYNGIS
Aditus laryngis
 Plica aryepiglottica
 Incisura interarytenoidea
Vestibulum laryngis
Rima vestibuli
 Plica vestibularis
Ventriculus laryngis
 Sacculus laryngis
Glottis
Rima glottidis
 Pars intermembranacea
 Pars intercartilaginea
 Plica vocalis

Plica interarytenoidea
Cavitas infraglottica
Tunica mucosa
 Gll. laryngeales
Membrana fibroelastica laryngis
 Membrana quadrangularis
  Ligamentum vestibulare
 Conus elasticus [membrana cricovocalis)
  Ligamentum vocale

TRACHEA

Pars cervicalis
Pars thoracica
Cartilagines tracheales
M. trachealis
Ligg. annularia [trachealia]
Paries membranaceus
Bifurcatio trachea
 Carina tracheae
Tunica mucosa

BRONCHI

Arbor bronchialis
Bronchus principalis [dexter et sinister]

BRONCHI LOBARES ET SEGMENTALES[59]
Bronchus lobaris superior dexter
 Bronchus segmentalis apicalis (B I)
 Bronchus segmentalis posterior (B II)
 Bronchus segmentalis anterior (B III)
Bronchus lobaris medius dexter
 Bronchus segmentalis lateralis (B IV)
 Bronchus segmentalis medialis (B V)
Bronchus lobaris inferior dexter[60]
 Bronchus segmentalis apicalis [superior]
  (B VI)
 Bronchus segmentalis basalis medialis
  [cardiacus] (B VII)
 Bronchus segmentalis basalis anterior (B
  VIII)
 B. segmentalis basalis lateralis (B IX)
 B. segmentalis basalis posterior (B X)

---

[59] At the request of many anatomists the segmental bronchi have been designated by Roman figures as useful abbreviations in clinical usage. If used verbally they will be presumably translated into vernaculars; otherwise, their Latin verbal equivalents (I = primus, II = secundus, etc.) must be used. Some members of the Committee regarded this as an unnecessary change.

[60] The Committee recommended deletion of the terms *Bronchus seg. subsuperior* [*subapicalis*] and of the corresponding segmental names, all of which were included *provisionally* in the third edition.

Bronchus lobaris superior sinister
  Bronchus segmentalis apicoposterior (B I + II)
  Bronchus segmentalis anterior (B III)
  Bronchus lingularis superior (B IV)
  Bronchus lingularis inferior (B V)
Bronchus lobaris inferior sinister[60]
  Bronchus segmentalis apicalis [superior] (B VI)
  Bronchus segmentalis basalis medialis [cardiacus] (B VII)
  Bronchus segmentalis basalis anterior (B VIII)
  Bronchus segmentalis basalis lateralis (B IX)
  Bronchus segmentalis basalis posterior (B X)
Rami bronchiales segmentorum
Tunica muscularis
Tela submucosa
Tunica mucosa
  Gll. bronchiales

PULMONES

PULMO DEXTER/SINISTER
Basis pulmonis
Apex pulmonis
Facies costalis
  Pars vertebralis
Facies mediastinalis
  Impressio cardiaca
Facies diaphragmatica
Facies interlobaris
Margo anterior
  Incisura cardiaca (pulmonis sinistri)
Margo inferior
Hilum pulmonis
Radix (Pediculus) pulmonis
Lingula pulmonis sinistri
Lobus superior
Lobus medius (pulmonis dextri)
Lobus inferior
Fissura obliqua
Fisura horizontalis (pulmonis dextri)

SEGMENTA BRONCHOPULMONALIA

Pulmo dexter, lobus superior
  Segmentum apicale (S I)
  Segmentum posterius (S II)
  Segmentum anterius (S III)

Pulmo dexter, lobus medius
  Segmentum laterale (S IV)
  Segmentum mediale (S V)

Pulmo dexter, lobus inferior
  Segmentum apicale [superius] (S VI)
  Segmentum basale mediale [cardiacum] (S VII)
  Segmentum basale anterius (S VIII)
  Segmentum basale laterale (S IX)
  Segmentum basale posterius (S X)

Pulmo sinister, lobus superior[61]
  Segmentum apicoposterius (S I + II)
  Segmentum anterius (S III)
  Segmentum lingulare superius (S IV)
  Segmentum lingulare inferius (S V)

Pulmo sinister, lobus inferior
  Segmentum apicale [superius] (S VI)
  Segmentum basale mediale [cardiacum] (S VII)
  Segmentum basale anterius (S VIII)
  Segmentum basale laterale (S IX)
  Segmentum basale posterius (S X)
Bronchioli
Bronchioli respiratorii
  Ductuli alveolares
  Sacculi alveolares
  Alveoli pulmonis

CAVITAS THORACIS

Regiones pleuropulmonales[62]
Fascia endothoracica
  Membrana suprapleuralis
  Fascia phrenicopleuralis
Cavitas pleuralis
Pleura
  Cupula pleurae
  Pleura visceralis [pulmonalis][63]

---

[61] The French Nomenclatural Commission suggested the term *Culmen pulmonis sinistri* for the part of the left *lobus superior* which is not lingular.

[62] These are new terms, = one-half the thoracic cavity, minus the mediastinum.

[63] *Visceralis* was preferred by the Committee, *pulmonalis* (third edition) being retained as a synonym.

Pleura parietalis
   Pleura mediastinalis
   Pleura costalis
   Pleura diaphragmatica
Recessus pleurales
   Recessus costodiaphragmaticus
   Recessus costomediastinalis
   Recessus phrenicomediastinalis
Lig. pulmonale
Mediastinum[64]
   Mediastinum superius
   Mediastinum inferius
     Mediastinum anterius
     Mediastinum medium
     Mediastinum posterius

APPARATUS UROGENITALIS
[SYSTEMA UROGENITALE]

ORGANA URINARIA[65]

REN
Margo lateralis
Margo medialis
   Hilum renale
   Sinus renalis
Facies anterior
Facies posterior
Extremitas superior
Extremitas inferior
Fascia renalis
   Corpus adiposum pararenale
Capsula adiposa
Capsula fibrosa

Segmenta renalia
   Segmentum superius
   Segmentum anterius superius
   Segmentum anterius inferius
   Segmentum inferius
   Segmentum posterius
Lobi renales
Cortex renalis

Pars convoluta
Pars radiata
Lobuli corticales
Medulla renalis
   Pyramides renales
     Basis pyramidis
     Papillae renales
     Area cribrosa
       Foramina papillaria
   Columnae renales

Arteriae renis [renales]
Arteriae interlobares
Arteriae arcuatae
   Arteriae interlobulares
     Arteriola glomerularis afferens
     [Vas afferens][66]
     Arteriola glomerularis efferens
     [Vas efferens][66]
Arteriolae rectae [Vasa recta]
Rami capsulares

Venae renis [renales]
Venae interlobares
Venae arcuatae
   Venae interlobulares
     Venulae rectae
Venulae stellatae

Pelvis renalis
Calices renales
   Calices renales majores
   Calices renales minores

URETER
Pars abdominalis
Pars pelvica
Tunica adventitia
Tunica muscularis
Tunica mucosa

VESICA URINARIA
Apex vesicae
Corpus vesicae
Fundus vesicae

[64] *Mediastinum inferius* has been added in this edition, but it is recognized that definitions of the subdivisions of the mediastinum are not identical in different countries. The schema shown here is current in Great Britain and the United States of America.

[65] *Organa urinaria* was suggested by the Soviet anatomists. This provides a more appropriate heading for a section containing more than the kidney, and it was preferred by the Committee to the *O. uropoietica* of the third edition.

[66] These terms have been altered to accord with *Nomina Histologica*. Their synonyms alone appeared in the third edition.

Cervix vesicae
Lig. umbilicale medianum
Tunica serosa
Tela subserosa
Tunica muscularis
  M. detrusor vesicae
  M. pubovesicalis
  M. rectovesicalis
  M. rectourethralis
Tela submucosa
Tunica mucosa
Trigonum vesicae
  Plica interureterica
  Ostium ureteris
  Ostium urethrae internum
Uvula vesicae

ORGANA GENITALIA MASCULINA INTERNA[67]

TESTIS [ORCHIS]
Extremitas superior
Extremitas inferior
Facies lateralis
Facies medialis
Margo anterior
Margo posterior
Tunica vaginalis testis
  Lamina parietalis
  Lamina visceralis
  Lig. epididymidis superius
  Lig. epididymidis inferius
  Sinus epididymidis
Tunica albuginea
Mediastinum testis
Septula testis
Lobuli testis
Parenchyma testis
Tubuli seminiferi contorti
Tubuli seminiferi recti
Rete testis
Ductuli efferentes testis

EPIDIDYMIS
Caput epididymidis
Corpus epididymidis
Cauda epididymidis

Lobuli epididymidis [Coni epididymidis]
Ductus epididymidis
Ductuli aberrantes
  (Ductulus aberrans superior)
  (Ductulus aberrans inferior)
  Appendix testis
  (Appendix epididymidis)
Paradidymis

DUCTUS DEFERENS
Ampulla ductus deferentis
  Diverticula ampullae
Tunica adventitia
Tunica muscularis
Tunica mucosa
Ductus ejaculatorius

VESICULA [GLANDULA] SEMINALIS
Tunica adventitia
Tunica muscularis
Tunica mucosa
Ductus excretorius

FUNICULUS SPERMATICUS (TUNICAE)[68]
Fascia spermatica externa
Musculus cremaster
Fascia cremasterica
Fascia spermatica interna
(Vestigium processus vaginalis)

PROSTATA
Basis prostatae
Apex prostatae
Facies anterior
Facies posterior
Facies inferolateralis
Lobus [dexter/sinister][69]
Isthmus prostatae
(Lobus medius)[69]
Capsula prostatica
Parenchyma
Ductuli prostatici
Substantia muscularis
M. puboprostaticus

GLANDULA BULBOURETHRALIS
  Ductus glandulae bulbourethralis

---

[67] *Interna* is added to this term. See footnote 70.
[68] Only the *coverings* of the spermatic cord are listed here.
[69] Definitions of these divisions of the prostate are subject to much controversy. Until disagreements are settled the Committee decided to make no nomenclatural change.

ORGANA GENITALIA MASCULINA EXTERNA[70]

PENIS
Radix penis
Corpus penis
Crus penis
Dorsum penis
Facies urethralis
Glans penis
   Corona glandis
   Septum glandis
   Collum glandis
Preputium penis
   Frenulum preputii
Raphe penis
Corpus cavernosum penis
Corpus spongiosum penis
Bulbus penis[71]
Tunica albuginea corporum cavernosorum
Tunica albuginea corporis spongiosi
Septum penis
Trabeculae corporum cavernosorum
Trabeculae corporis spongiosi
Cavernae corporum cavernosorum
Cavernae corporis spongiosi
Arteriae helicinae
Venae cavernosae
Fascia penis superficialis
Fascia penis profunda
Glandulae preputiales

URETHRA MASCULINA
Pars prostatica
   Crista urethralis
   Colliculus seminalis
   Utriculus prostaticus
   Sinus prostaticus[72]
Pars membranacea
Pars spongiosa
   Fossa navicularis urethrae
   (Valvula fossae navicularis)
Ostium urethrae externum
Lacunae urethrales

Gll. urethrales
Ductus [Canales] paraurethrales

SCROTUM
Raphe scroti (scrotalis)
Septum scroti (scrotalis)
Tunica dartos
   M. dartos

ORGANA GENITALIA FEMININA INTERNA

OVARIUM
Hilum ovarii
Facies medialis
Facies lateralis
Margo liber
Margo mesovaricus
Extremitas tubaria (tubale)
Extremitas uterina
Tunica albuginea
Stroma ovarii
Cortex ovarii
Medulla ovarii
Folliculi ovarici primarii
Folliculi ovarici vesiculosi[73]
Corpus luteum
Corpus albicans
Ligamentum ovarii proprium

TUBA UTERINA [SALPINX][74]
Ostium abdominale tubae uterinae
Infundibulum tubae uterinae
Fimbriae tubae
   Fimbria ovarica
Ampulla tubae uterinae
Isthmus tubae uterinae
Pars uterina
Ostium uterinum tubae
Tunica serosa
Tela subserosa
Tunica muscularis

---

[70] This title was *Partes Genitales Masculinae Externae* in the third edition. See footnote 67.
[71] This is the enlarged proximal part of the *Corpus spongiosum*. The alternative *Bulbus corporis spongiosi* is regarded as more suitable by a minority of the Committee.
[72] This is the recess between the *Colliculus seminalis* and the urethral wall into which the *Ductuli prostatici* discharge.
[73] Now that a *Nomina Histologica* is available it is possible to delete many histological terms from *Nomina Anatomica*, as has been done here regarding the microscopic details of the ovary, including the ovum.
[74] *Salpinx* has been added as a synonym, since most clinical expressions use this stem.

Tunica mucosa
Plicae tubariae (tubales)

UTERUS
Corpus uteri
Fundus uteri
Cornu uteri [dextrum/sinistrum]
Margo uteri [dexter/sinister]
Facies intestinalis
Cavitas uteri
Facies vesicalis
Isthmus uteri
Cervix uteri
   Portio supravaginalis (cervicis)
   Portio vaginalis (cervicis)
Ostium uteri
   Labium anterius
   Labium posterius
Canalis cervicis uteri
   Plicae palmatae
   Glandulae cervicales [uteri]
Parametrium
Paracervix
Tunica serosa [Perimetrium]
Tela subserosa
Tunica muscularis [Myometrium]
   M. recto-uterinus
Tunica mucosa [Endometrium]
   Gll. uterinae
Lig. teres uteri

VAGINA
Fornix vaginae
   Pars anterior
   Pars posterior
   Pars lateralis
Paries anterior
Paries posterior
Hymen
   Carunculae hymenales
Tunica muscularis
Tunica mucosa
   Rugae vaginales
   Columnae rugarum
      Columna rugarum anterior
      Columna rugarum posterior
      Carina urethralis vaginae
Tunica spongiosa

Epoöphoron
Ductus epoöphorontis longitudinalis
   Ductuli transversi
Appendices vesiculosae

Paroöphoron
(Ductus deferens vestigialis)

ORGANA GENITALIA FEMININA EXTERNA[75]

PUDENDUM FEMININUM
Mons pubis
Labium majus pudendi
   Commissura labiorum anterior
   Commissura labiorum posterior
   Rima pudendi
Labium minus pudendi
   Frenulum labiorum pudendi
Vestibulum vaginae
   Fossa vestibuli vaginae
Bulbus vestibuli
   Pars intermedia (Commissura) bulborum
Ostium vaginae
Gll. vestibulares minores
Gl. vestibularis major

Clitoris
Crus clitoridis
Corpus clitoridis
Glans clitoridis
   Frenulum clitoridis
Preputium clitoridis
Corpus cavernosum clitoridis [dextrum/
   sinistrum]
Septum corporum cavernosorum
Fascia clitoridis

URETHRA FEMININA
Ostium urethrae externum
Tunica muscularis
Tunica spongiosa
Tunica mucosa
   Gll. urethrales
   Lacunae urethrales
   (Ductus paraurethrales)
Crista urethalis

---

[75] See note 70.

PERINEUM

Raphe perinealis
Musculi perinei [perineales]
Centrum tendineum perinei[76]

DIAPHRAGMA PELVIS (*see* page A 29)
M. levator ani
  M. pubococcygeus
  M. levator prostatae [m. pubovaginalis]
  M. puborectalis
  M. iliococcygeus
  (Arcus tendineus m. levatoris ani)
  Lig. anococcygeum
M. coccygeus
M. sphincter ani externus
  Pars subcutanea
  Pars superficialis
  Pars profunda
Fascia pelvis
  Fascia pelvis [pelvica] parietalis
    Fascia obturatoria
  Fascia pelvis visceralis
    Fascia prostatae
    Fascia peritoneoperinealis
      Septum rectovesicale
      Septum rectovaginale
  Fascia diaphragmatis pelvis superior
  Arcus tendineus fasciae pelvis
  Lig. puboprostaticum
        [pubovesicale]
  Fascia diaphragmatis pelvis
      inferior

SPATIUM PERINEI PROFUNDUM[77,79]
M. transversus perinei profundus
M. sphincter urethrae
M. compressor urethrae
M. sphincter urethrovaginalis
Membrana perinei
  (Fascia diaphragmatis urogenitalis
      inferior)[79]
Lig. transversum perinei

SPATIUM PERINEI SUPERFICIALE[78]
M. transversus perinei superficialis
M. ischiocavernosus
M. bulbospongiosus
Fascia perinei superficialis

FOSSA ISCHIOANALIS
Corpus adiposum fossae ischioanalis
Canalis pudendalis

PERITONEUM

Cavitas peritonealis
Spatium extraperitoneale
  Spatium retroperitoneale
  Spatium retropubicum
  Fascia extraperitonealis
  Organum extraperitoneale
Peritoneum parietale
  Tunica serosa
  Tela subserosa
Peritoneum viscerale
  Tunica serosa
  Tela subserosa
Foramen omentale [epiploicum][80]
Bursa omentalis
  Vestibulum bursae omentalis
  Recessus superior omentalis
  Recessus inferior omentalis
  Recessus splenicus [lienalis]
  Plica gastropancreatica
  Plica hepatopancreatica
Mesenterium
  Radix mesenterii
Mesocolon
  Mesocolon transversum
  Mesocolon ascendens
  Mesocolon descendens
  Mesocolon sigmoideum
  Mesoappendix

OMENTUM MINUS
Lig. hepatogastricum

---

[76] This is the perineal body of gynaecological usage. Perhaps the synonym *Corpus perinealis* should be added.

[77] This is the *region*, rather than space, superior to the *Membrana perinei*.

[78] The region between the *Membrana perinei* and the *Fascia perinei superficialis*.

[79] The alterations in this section reflect the fact that the commonly described and illustrated flat sandwich of urogenital diaphragm does not actually exist—there is no superior fascia separating the sphincter urethrae muscle from the prostate gland and this muscle rises on the prostate almost to the base of the bladder. The deep perineal space is thus a region and its superior limit may be considered to be the dense endopelvic fascia of the floor of the pelvis.

[80] Epiploön and its derivatives are today unfamiliar.

Lig. hepatoduodenale
(Lig. hepatocolicum)

OMENTUM MAJUS
Lig. gastrophrenicum
Lig. gastrosplenicum
  [gastrolienale]
Lig. gastrocolicum
Lig. splenorenale [lienorenale,
  phrenicosplenicum]

LIGAMENTA HEPATIS
Lig. coronarium
  Lig. falciforme (hepatis)
  Lig. triangulare dextrum
  Lig. triangulare sinistrum
  Lig. hepatorenale

PLICAE ET FOSSAE [RECESSUS]
Fascia retinens rostralis
Plica duodenalis superior [Plica
  duodenojejunalis]
Recessus duodenalis superior
Plica duodenalis inferior [Plica
  duodenomesocolica]
Recessus duodenalis inferior
(Plica paraduodenalis)
(Recessus paraduodenalis)
(Recessus retroduodenalis)
Recessus intersigmoideus
Recessus ileocaecalis [cec-] superior
Plica caecalis [cec-] vascularis
Recessus ileocaecalis [cec-] inferior
Plica ileocaecalis [cec-]
Recessus retrocaecalis [-cec-]
Plicae caecales [cec-]
Sulci paracolici
Recessus subphrenici
Recessus subhepatici
Recessus hepatorenalis

PERITONEUM PARIETALE ANTERIUS
Plica umbilicalis mediana
Fossa supravesicalis
Plica umbilicalis medialis[81]
Fossa inguinalis medialis
Trigonum inguinale
Plica umbilicalis lateralis[82]

Fossa inguinalis lateralis
Plica vesicalis transversa
Fossa paravesicalis

PERITONEUM UROGENITALE
Lig. latum uteri
  Mesometrium
  Mesosalpinx
  Mesovarium
Lig. suspensorium ovarii
Fossa ovarica
Plica rectouterina
Excavatio rectouterina
Excavatio vesicouterina
Excavatio rectovesicalis
Fossa paravesicalis

GLANDULAE ENDOCRINAE

GLANDULA THYROIDEA

Lobus [dexter/sinister]
Isthmus glandulae thyroideae
(Lobus pyramidalis)
(Gll. thyroideae accessoriae)
Capsula fibrosa
Stroma
Parenchyma
Lobuli

GLANDULA PARATHYROIDEA
   SUPERIOR

GLANDULA PARATHYROIDEA
   INFERIOR

HYPOPHYSIS [GLANDULA
   PITUITARIA (see page A70)

Adenohypophysis [Lobus anterior]
  Pars tuberalis
  Pars intermedia[83]
  Pars distalis
Neurohypophysis [Lobus posterior]

---

[81] Associated with the obliterated umbilical artery.
[82] Associated with the inferior epigastric artery.
[83] Uncertainty still exists as to the limits between the anterior and posterior lobes.

Infundibulum
Lobus nervosus

CORPUS PINEALE [GLANDULA PINEALIS]
(*see* page A68)

THYMUS

Lobus [dexter/sinister]
(Noduli thymici accessorii)
Lobuli thymi
Cortex thymi
Medulla thymi

GLANDULA SUPRARENALIS
[ADRENALIS]

Facies anterior
Facies posterior
Facies renalis
Margo superior
Margo medialis
Hilum
    Vena centralis
Cortex
Medulla
(Gll. suprarenales accessoriae)

ANGIOLOGIA

Anastomosis arteriolovenularis
    [arteriovenosa]
Arteria
Arteria nutricia [nutriens]
Arteriola
Circulus arteriosus
Circulus vasculosus
Cisterna
Haema [Hema]
Lympha
Nervi vasorum
Nodus lymphaticus [Lymphonodus][84]

Nodulus [folliculus] lymphaticus[85]
Plexus lymphaticus
Plexus vasculosus
Plexus venosus
Rete arteriosum
Rete mirabile
Rete vasculosum articulare
Rete venosum
Sinus venosus
Tunica externa
Tunica intima
Tunica media
Valva
Valvula lymphatica
Valvula venosa
Vas anastomoticum
Vas capillare
Vas collaterale
Vas lymphaticum
Vas sinusoideum
Vasa vasorum
Vena
Vena comitans
Vena cutanea
Vena emissaria
Vena profunda
Vena superficialis
Venula

PERICARDIUM

Pericardium fibrosum
    Ligamenta sternopericardiaca
Pericardium serosum
    Lamina parietalis
    Lamina visceralis (Epicardium)
Cavitas pericardialis
    Sinus transversus pericardii
    Sinus obliquus pericardii

COR

Basis cordis
Facies sternocostalis (anterior)
Facies diaphragmatica (inferior)

---

[84] *Lymphonodus* is favoured by many anatomists and histologists and is hence included as an official alternative.
[85] *Folliculus* is retained as an official alternative, despite strong objections from histologists (based on the lack of a central cavity in such primary lymphatic nodules), because of its wide use, at present, by immunologists, pathologists, and others.

Facies pulmonalis
Margo dexter[86]
Apex cordis
   Incisura apicis cordis
Sulcus interventricularis anterior
Sulcus interventricularis posterior[87]
Sulcus coronarius
Ventriculus cordis
Septum interventriculare
   Pars muscularis
   Pars membranacea
Septum atrioventriculare
Atrium cordis
   Auricula atrialis
Septum interatriale
Trabeculae carneae
Vortex cordis
Musculi papillares
Chordae tendineae
Trigonum fibrosum dextrum
Trigonum fibrosum sinistrum
Annuli fibrosi
Tendo infundibuli[88]

MYOCARDIUM
Systema conducens cordis
   Nodus sinuatrialis
   Nodus atrioventricularis
   Fasciculus atrioventricularis
   Truncus
   Crus dextrum
   Crus sinistrum
   Rami subendocardiales

ENDOCARDIUM

ATRIUM DEXTRUM

Auricula dextra
Crista terminalis
(Foramen ovale)
Foramina venarum minimarum
Fossa ovalis

Limbus fossae ovalis
Musculi pectinati
Ostium sinus coronarii
Ostium venae cavae inferioris
Ostium venae cavae superioris
Sinus venarum cavarum
Sulcus terminalis
Tuberculum intervenosum
Valvula venae cavae inferioris
Valvula sinus coronarii

VENTRICULUS DEXTER

Ostium atrioventriculare dextrum
Valva atrioventricularis dextra[89]
   Cuspis anterior
   Cuspis posterior
   Cuspis septalis
Crista supraventricularis
Conus arteriosus
Ostium trunci pulmonalis[90]
Valva trunci pulmonalis
   Valvula semilunaris anterior[91]
   Valvula semilunaris dextra
   Valvula semilunaris sinistra
      Noduli valvularum semilunarium
      Lunulae valvularum semilunarium
Musculus papillaris anterior
Musculus papillaris posterior
Trabecula septomarginalis[92]
Trabeculae carneae

ATRIUM SINISTRUM

Auricula sinistra
Musculi pectinati
Ostia venarum pulmonalium
Valvula foraminis ovalis (Falx septi)

VENTRICULUS SINISTER

Ostium atrioventriculare sinistrum

---

[86] *Margo* is unfortunate, because this is a "facies," except in radiograms and two-dimensional illustrations.
[87] This *Sulcus* is, of course, inferior in position, and not posterior.
[88] This is the vestige of the *Septum spirale* of the embryonic heart.
[89] The term *Valva* is to define the entire valvular mechanism. The terms *Valvula* and *Cuspis* are used almost as synonyms, but the latter possesses *Chordae tendineae*.
[90] *Truncus* is introduced to define the *Arteria pulmonalis* before its bifurcation.
[91] The term Valva is to define the entire valvular mechanism. The terms *Valvula* and *Cuspis* are used almost as synonyms, but the latter possesses *Chordae tendineae*.
[92] This is the "moderator band."

Valva atrioventricularis sinistra[91]
  Cuspis anterior
  Cuspis posterior
  Cuspides commissurales
Musculus papillaris anterior
Musculus papillaris posterior
Ostium aortae
Trabeculae carneae
Valva aortae
  Valvula semilunaris dextra[91]
  Valvula semilunaris posterior
  Valvula semilunaris sinistra
    Noduli valvularum semilunarium
    Lunulae valvularum semilunarium

ARTERIAE

TRUNCUS PULMONALIS[93,94]

Sinus trunci pulmonalis
Bifurcatio trunci pulmonalis

ARTERIA PULMONALIS DEXTRA
Rami lobi superioris
  Ramus apicalis
  Ramus anterior ascendens
  Ramus anterior descendens
  Ramus posterior ascendens
  Ramus posterior descendens
Rami lobi medii
  Ramus medialis
  Ramus lateralis
Rami lobi inferioris
  Ramus superior lobi inferioris
  Pars basalis
    Ramus basalis anterior
    Ramus basalis lateralis
    Ramus basalis medialis
    Ramus basalis posterior

ARTERIA PULMONALIS SINISTRA
Ligamentum arteriosum (*Ductus arteriosus*)
Rami lobi superioris
  Ramus apicalis
  Ramus anterior ascendens
  Ramus anterior descendens
  Ramus posterior
  Ramus lingularis
    Ramus lingularis inferior

    Ramus lingularis superior
Rami lobi inferioris
  Ramus superior lobi inferioris
  Pars basalis
  Ramus basalis anterior
  Ramus basalis lateralis
  Ramus basalis medialis
  Ramus basalis posterior

AORTA

PARS ASCENDENS

AORTAE
[AORTA ASCENDENS]
Bulbus aortae
Sinus aortae

Arteria coronaria dextra
  Ramus coni arteriosi
  Ramus nodi sinuatrialis
  Rami atriales
  Ramus marginalis dexter
  Ramus atrialis intermedius
  Ramus interventricularis posterior
    Rami interventriculares septales
  Ramus nodi atrioventricularis
  (Ramus posterolateralis dexter)

Arteria coronaria sinistra
  Ramus interventricularis anterior
  Ramus coni arteriosi
  Ramus lateralis
  Rami interventriculares septales
  Ramus circumflexus
    Ramus atrialis anastomoticus
    Rami atrioventriculares
    Ramus marginalis sinister
    Ramus atrialis intermedius
    Ramus posterior ventriculi sinistri
    (Ramus nodi sinuatrialis)
    (Ramus nodi atrioventricularis)
    Rami atriales

ARCUS AORTAE

Isthmus aortae
Corpora para-aortica [Glomera aorticae]

---

[93] *Truncus* is introduced to define the *Arteria pulmonalis* before its bifurcation.
[94] This classification is based on the work of E. A. Boyden, *Segmental Anatomy of the Lungs*, McGraw Hill, 1955.

TRUNCUS BRACHIOCEPHALICUS
(Arteria thyreoidea ima)

ARTERIA CAROTIS COMMUNIS
Glomus caroticum
Sinus caroticus
Bifurcatio carotidis

ARTERIA CAROTIS EXTERNA

Arteria thyreoidea superior
  Ramus infrahyoideus
  Ramus sternocleidomastoideus
  A. laryngea superior
  Ramus cricothyroideus
  Ramus glandularis anterior
  Ramus glandularis posterior
  Ramus glandularis lateralis

Arteria pharyngea ascendens
  A. meningea posterior
  Rami pharyngeales
  A. tympanica inferior

Arteria lingualis
  Ramus suprahyoideus
  Rami dorsales linguae
  A. sublingualis
  A. profunda linguae

Arteria facialis
  A. palatina ascendens
  Ramus tonsillaris
  A. submentalis
  Rami glandulares
  A. labialis inferior
  A. labialis superior
    Ramus septi nasi
  Ramus lateralis nasi
  A. angularis

(Truncus linguofacialis)[95]

Arteria occipitalis
  Ramus mastoideus
  Ramus auricularis
  Rami sternocleidomastoidei
  Rami occipitales
  (Ramus meningeus)
  Ramus descendens

Arteria auricularis posterior
  A. stylomastoidea
  A. tympanica posterior
  Rami mastoidei
  (Ramus stapedialis)
  Ramus auricularis
  Ramus occipitalis
  Ramus parotideus

Arteria temporalis superficialis
  Ramus parotideus
  A. transversa facialis
  Rami auriculares anteriores
  A. zygomatico-orbitalis
  A. temporalis media
  Ramus frontalis
  Ramus parietalis

Arteria maxillaris
  A. auricularis profunda
  A. tympanica anterior
  A. alveolaris inferior
    Rami dentales
    Rami peridentales
    Ramus mentalis
    Ramus mylohyoideus
  A. meningea media
    Ramus frontalis
    Ramus parietalis
    Ramus orbitalis
    Ramus petrosus
    A. tympanica superior
    Ramus anastomaticus (cum A. lacrimali)
  A. pterygomeningea
  A. masseterica
  A. temporalis profunda anterior
  A. temporalis profunda posterior
  Rami pterygoidei
  A. buccalis
  A. alveolaris superior posterior
    Rami dentales
    Rami peridentales
  A. infraorbitalis
    Aa. alveolares superiores anteriores
    Rami dentales
    Rami peridentales
  A. canalis pterygoidei
    Ramus pharyngeus
  A. palatina descendens

[95]   The lingual and facial arteries may branch from a common trunk.

A. palatina major
Aa. palatinae minores
Ramus pharyngeus
A. sphenopalatina
Aa. nasales posteriores laterales
Rami septales posteriores

ARTERIA CAROTIS INTERNA

Pars cervicalis
Sinus caroticus

Pars petrosa
Aa. caroticotympanicae
A. canalis pterygoidei

Pars cavernosa
Ramus basalis tentorii
Ramus marginalis tentorii
Ramus meningeus
Ramus sinus cavernosi
A. hypophysialis inferior
Ramus ganglionis trigemini
Rami nervorum

Pars cerebralis
A. hypophysialis superior
Ramus clivi

Arteria ophthalmica
A. centralis retinae
A. lacrimalis
Ramus anastomoticus (cum A. meningea media)
Aa. palpebrales laterales
Ramus meningeus recurrens
Aa. ciliares posteriores breves
Aa. ciliares posteriores longae
Aa. musculares
Aa. ciliares anteriores
Aa. conjunctivales anteriores
Aa. episclerales
A. supraorbitalis
A. ethmoidalis anterior
Ramus meningeus anterior
Rami septales anteriores
Rami nasales anteriores laterales
A. ethmoidalis posterior
Aa. palpebrales mediales
Aa. conjunctivales posteriores

Arcus palpebralis inferior
Arcus palpebralis superior
A. supratrochlearis
A. dorsalis nasi [A. nasi externa]

Arteria communicans posterior

Arteria choroidea anterior
Rami choroidei ventriculi lateralis
Rami choroidei ventriculi tertii
Rami substantiae perforatae anterioris
Rami tractus optici
Rami corporis geniculati lateralis
Rami capsulae internae
Rami globi pallidi
Rami caudae nuclei caudati
Rami tuberis cinerei
Rami nucleorum hypothalamicorum
Rami substantiae nigrae
Rami nuclei rubris
Rami corporis amygdaloidei

Arteria cerebri anterior[96]
Pars precommunicalis
Aa. centrales anteromediales [Aa. thalamostriatae anteromediales]
A. centralis brevis
A. centralis longa [A. recurrens]
A. communicans anterior
Pars postcommunicalis [A. pericallosa]
A. frontobasalis medialis [Ramus orbitofrontalis medialis]
A. callosomarginalis
Ramus frontalis anteromedialis
Ramus frontalis mediomedialis
Ramus frontalis posteromedialis
Ramus cingularis
A. paracentralis
A. precunealis
A. parieto-occipitalis

Arteria cerebri media
Pars sphenoidalis
Aa. centrales anterolaterales [Aa. thalamostriatae anterolaterales]
Rami laterales
Rami mediales
Pars insularis
Aa. insulares

[96] The list of cerebral arteries has been amplified, at the insistence of radiologists and others and following a resolution of the I.A.N.C. in Leningrad, 1970.

A 51

A. frontobasalis lateralis [Ramus
  orbitofrontalis lateralis]
A. temporalis anterior
A. temporalis media
A. temporalis posterior
Pars terminalis [Pars corticalis]
A. sulci centralis
A. sulci precentralis
A. sulci postcentralis
Aa. parietales anterior et posterior
A. gyri angularis

ARTERIA SUBCLAVIA

Arteria vertebralis
Pars prevertebralis
Pars transversaria [cervicalis]
Rami spinales [radiculares]
Rami musculares
Pars atlantica [atlantis]
Pars intracranialis
Rami meningei
A. spinalis anterior
A. inferior posterior cerebelli
Ramus choroideus ventriculi quarti
Ramus tonsillae cerebelli
Rami medullares mediales et
  laterales
  [Rami ad medullam oblongatam]

Arteria basilaris
A. inferior anterior cerebelli
A. spinalis posterior
A. labyrinthi [Ramus meatus acustici
  interni]
Aa. pontis
Aa. mesencephalicae
A. superior cerebelli

Arteria cerebri posterior
Pars precommunicalis
Aa. centrales posteromediales
Pars postcommunicalis
Aa. centrales posterolaterales
Rami thalamici
Rami choroidei posteriores mediales
Rami choroidei posteriores laterales
Rami pedunculares
Pars terminalis [corticalis]
A. occipitalis lateralis

Rami temporales anteriores
Rami temporales [intermedii
  mediales]
Rami temporales posteriores
A. occipitalis medialis
Ramus corporis callosi dorsalis
Ramus parietalis
Ramus parieto-occipitalis
Ramus calcarinus
Ramus occipitotemporalis

Circulus arteriosus cerebri
A. carotis interna
A. cerebri anterior
A. communicans anterior
Aa. centrales anteromediales
A. cerebri media
A. communicans posterior
Ramus chiasmaticus
Ramus nervi oculomotorii
Ramus thalamicus
Ramus hypothalamicus
Ramus caudae nuclei caudati
A. cerebri posterior

Arteria thoracica interna
Rami mediastinales
Rami thymici
(Rami bronchiales)
(Rami tracheales)
A. pericardiacophrenica
Rami sternales
Rami perforantes
Rami mammarii mediales
(Ramus costalis lateralis)
Rami intercostales anteriores
A. musculophrenica
A. epigastrica superior

Truncus thyrocervicalis
A. thyroidea inferior
A. laryngea inferior
Rami glandulares
Rami pharyngeales
Rami oesophageales [eso—]
Rami tracheales
A. cervicalis ascendens
Rami spinales
A. suprascapularis
Ramus acromialis

A 52

A. transversa cervicis[97]
  Ramus superficialis [A. cervicalis
    superficialis]
    Ramus ascendens
    Ramus descendens
  Ramus profundus [A. dorsalis scapulae]
A. dorsalis scapulae [A. scapularis
    dorsalis][98]

Truncus costocervicalis
  A. cervicalis profunda
  A. intercostalis suprema
    A. intercostalis posterior prima
    A. intercostalis posterior secunda
    Rami dorsales

ARTERIA AXILLARIS
Rami subscapulares
A. thoracica superior
A. thoracoacromialis
  Ramus acromialis
  Ramus clavicularis
  Ramus deltoideus
  Rami pectorales
A. thoracica lateralis
  Rami mammarii laterales
A. subscapularis
  A. thoracodorsalis
  A. circumflexa scapulae
A. circumflexa anterior humeri
A. circumflexa posterior humeri

ARTERIA BRACHIALIS
(A. brachialis superficialis)[99]
A. profunda brachii
    Aa. nutriciae [nutrientes] humeri
  Ramus deltoideus
  A. collateralis media
  A. collateralis radialis
A. collateralis ulnaris superior
A. collateralis ulnaris inferior

Arteria radialis
A. recurrens radialis
Ramus carpalis palmaris

Ramus palmaris superficialis
Ramus carpalis dorsalis
  Rete carpale dorsale
    Aa. metacarpales dorsales
    Aa. digitales dorsales
A. princeps pollicis
A. radialis indicis
Arcus palmaris profundus
    Aa. metacarpales palmares
    Rami perforantes

Arteria ulnaris
A. recurrens ulnaris
  Ramus anterior
  Ramus posterior
Rete articulare cubiti
A. interossea communis
    A. interossea anterior
      A. comitans nervi mediani
    A. interossea posterior
      A. interossea recurrens
Ramus carpalis dorsalis
Ramus carpalis palmaris
Ramus palmaris profundus
Arcus palmaris superficialis
Aa. digitales palmares communes
    Aa. digitales palmares propriae

PARS DESCENDENS AORTAE
  [AORTA DESCENDENS]

PARS THORACICA AORTAE
  [AORTA THORACICA]
Rami bronchiales
Rami oesophageales [eso—]
Rami pericardiaci
Rami mediastinales
Aa. phrenicae superiores
Aa. intercostales posteriores (tertia usque ad
    undecimam)[100]
  Ramus dorsalis
    Ramus cutaneus medialis
    Ramus cutaneus lateralis
    Rami spinales
  Ramus collateralis

---

[97] The alternative *Colli* has been used but very few structures are in fact named after the *Collum*. This
    artery may be derived directly from the third part of the subclavian.
[98] The *Arteria dorsalis scapulae* may be derived directly from the subclavian or from the *a. transversa
    cervicis*.
[99] A variant of the *Arteria brachialis*.
[100] This includes the 3rd to 11th paired arteries, the 1st and 2nd being branches of the *Arteria intercostalis
    suprema* and the 12th is the *A. subcostalis*.

A 53

Ramus cutaneus lateralis
Rami mammarii laterales
A. subcostalis
Ramus dorsalis
Ramus spinalis

[PARS ABDOMINALIS AORTAE]
[AORTA ABDOMINALIS]

Arteria phrenica inferior
Aa. suprarenales superiores

Arteriae lumbales
Ramus dorsalis
Ramus spinalis

Arteria sacralis mediana
Aa. lumbales imae
Rami sacrales laterales
Glomus coccygeum[101]

Truncus coeliacus [cel—]
A. gastrica sinistra
Rami oesophageales [eso—]
A. hepatica communis[102]
A. hepatica propria
A. gastrica dextra
Ramus dexter
A. cystica
A. lobi caudati
A. segmenti anterioris
A. segmenti posterioris
Ramus sinister
A. lobi caudati
A. segmenti medialis
A. segmenti lateralis
Ramus intermedius[103]
A. gastroduodenalis
(A. supraduodenalis)
A. pancreaticoduodenalis superior
posterior
Rami pancreatici
Rami duodenales
Aa. retroduodenales
A. gastro-omentalis [epiploica] dexter
Rami gastrici
Rami omentales [epiploici]

A. pancreaticoduodenalis superior
anterior
Rami pancreatici
Rami duodenales
A. splenica [lienalis]
Rami pancreatici
A. pancreatica dorsalis
A. pancreatica inferior
A. pancreatica magna
A. caudae pancreatis
A. gastro-omentalis [epiploica] sinistra
Rami gastrici
Rami omentales
Aa. gastricae breves
Rami splenici
A. gastrica posterior

Arteria mesenterica superior
A. pancreaticoduodenalis inferior
Ramus anterior
Ramus posterior
Aa. jejunales
Aa. ileales
A. ileocolica
A. caecalis [cec-]anterior
A. caecalis [cec-] posterior
A. appendicularis
Ramus ilealis
Ramus colicus
A. colica dextra
A. colica media

Arteria mesenterica inferior
A. colica sinistra
Aa. sigmoideae
A. rectalis superior

Arteria suprarenalis media

Arteria renalis
A. suprarenalis inferior
Ramus anterior
A. segmenti superioris
A. segmenti anterioris superioris
A. segmenti anterioris inferioris
A. segmenti inferioris

---

[101] This structure is controversial in nature, but is included here for convenience. It is most probably a glomus.

[102] Not all the segmental branches named by various authorities have been included.

[103] This branch is the main supply to the quadrate lobe and can arise from the left or right hepatic ramus. (This branch could be omitted if it is among the *Arteria segmenti medialis*).

Ramus posterior
  A. segmenti posterioris
Rami ureterici

Arteria testicularis
  Rami ureterici
  Rami epididymales

Arteria ovarica
  Rami ureterici
  Rami tubarii [tubales]

BIFURCATIO AORTAE

ARTERIA ILIACA COMMUNIS

ARTERIA ILIACA INTERNA

Arteria iliolumbalis
  Ramus lumbalis
    Ramus spinalis
  Ramus iliacus

Arteriae sacrales laterales
  Rami spinales

Arteria obturatoria
  Ramus pubicus
  Ramus acetabularis
  Ramus anterior
  Ramus posterior

Arteria glutea superior
  Ramus superficialis
  Ramus profundus
    Ramus superior
    Ramus inferior

Arteria glutea inferior
  A. comitans nervi ischiadici

Arteria umbilicalis
  Pars patens
    A. ductus deferentis
      Rami ureterici
    Aa. vesicales superiores
  Pars occlusa
  Ligamentum umbilicale mediale

Arteria vesicalis inferior
  Rami prostatici

Arteria uterina
  Rami helicini
  Rami vaginales [Aa. azygoi vaginae]
  Ramus ovaricus
  Ramus tubarius [tubalis]

Arteria vaginalis

Arteria rectalis media
  (Rami vaginales)

Arteria pudenda interna
  A. rectalis inferior
  A. perinealis
  Rami scrotales/labiales posteriores
  A. urethralis
  A. bulbi penis
  A. bulbi vestibuli [vaginae]
  A. dorsalis penis/clitoridis
  A. profunda penis/clitoridis

ARTERIA ILIACA EXTERNA

Arteria epigastrica inferior
  Ramus pubicus
    Ramus obturatorius
    (A. obturatoria accessoria)
  A. cremasterica
  A. ligamenti teretis uteri

Arteria circumflexa iliaca profunda
  Ramus ascendens

ARTERIA FEMORALIS

Arteria epigastrica superficialis

Arteria circumflexa iliaca superficialis

Arteriae pudendae externae
  Rami scrotales labiales anteriores
  Rami inguinales

Arteria descendens genicularis
  Ramus saphenus
  Rami articulares

ARTERIA PROFUNDA FEMORIS[104]

---

[104] The arrangement of the branches of this artery is subject to much variation. The arrangement adopted here is merely the most common. The origin of the circumflex femoral branches is particularly variable.

A. circumflexa femoris medialis
    Ramus profundus
    Ramus ascendens
    Ramus transversus
    Ramus acetabularis
A. circumflexa femoris lateralis
    Ramus ascendens
    Ramus descendens
    Ramus transversus
Aa. perforantes
    Aa. nutriciae [nutrientes] femoris

ARTERIA POPLITEA
A. superior lateralis genus
A. superior medialis genus
A. media genus
Aa. surales
A. inferior lateralis genus
A. inferior medialis genus
Rete articulare genus
Rete patellae

Arteria tibialis anterior
    A. recurrens tibialis anterior
    (A. recurrens tibialis posterior)
    A. malleolaris anterior lateralis
    A. malleolaris anterior medialis
    Rete malleolare laterale
Arteria dorsalis pedis
    A. tarsalis lateralis
    Aa. tarsales mediales
    (A. arcuata)
    Aa. metatarsales dorsales
    Aa. digitales dorsales
    A. plantaris profundus

Arteria tibialis posterior
    Ramus circumflexus fibularis
    Rami malleolares mediales
    Rami calcanei
    A. nutricia [nutriens] tibiae

Arteria plantaris medialis
    Ramus profundus
    Ramus superficialis

Arteria plantaris lateralis
    Arcus plantaris profundus
      Aa. metatarsales plantares
        Rami perforantes
        Aa. digitales plantares communes
        Aa. digitales plantares propriae

(Arcus plantaris superficialis)
Arteria fibularis [peronea]
    Ramus perforans
    Ramus communicans
    Rami malleolares laterales
      Rami calcanei
    Rete calcaneum
    A. nutricia [nutriens] fibulae

VENAE

VENAE PULMONALES

VENAE PULMONALES DEXTRAE

Vena pulmonalis dextra superior
    Ramus apicalis
      Pars intrasegmentalis
      Pars intersegmentalis
    Ramus anterior
      Pars intrasegmentalis
      Pars intersegmentalis
    Ramus posterior
      Pars infralobaris
      Pars intralobaris (intersegmentalis)
    Ramus lobi medii
      Pars lateralis
      Pars medialis

Vena pulmonalis dextra inferior
    Ramus superior
      Pars intrasegmentalis
      Pars intersegmentalis
    V. basalis communis
      V. basalis superior
        Radix basalis anterior
          Pars intrasegmentalis
          Pars intersegmentalis
      V. basalis inferior

VENAE PULMONALES SINISTRAE

Vena pulmonalis sinistra superior
    Ramus apicoposterior
      Pars intrasegmentalis
      Pars intersegmentalis
    Ramus anterior
      Pars intrasegmentalis
      Pars intersegmentalis
    Ramus lingularis

Pars superior
Pars inferior

Vena pulmonalis sinistra inferior
 Ramus superior
  Pars intrasegmentalis
  Pars intersegmentalis
 V. basalis communis
 V. basalis superior
  Ramus basalis anterior
   Pars intrasegmentalis
   Pars intersegmentalis
 V. basalis inferior

VENAE CORDIS

Sinus coronarius
V. cardiaca magna
V. posterior ventriculi sinistri
V. obliqua atrii sinistri
Plica v. cavae sinistrae
V. cardiaca media
V. cardiaca parva
Vv. cardiacae anteriores
Vv. cardiacae minimae
Vv. atriales
Vv. ventriculares
Vv. atrioventriculares

VENA CAVA SUPERIOR

VENA BRACHIOCEPHALICA
(DEXTRA/SINISTRA)

V. thyroidea inferior
Plexus thyroideus impar
 V. laryngea inferior
Vv. thymicae
Vv. pericardiacae
Vv. pericardiacophrenicae
Vv. mediastinales

Vv. bronchiales
Vv. tracheales
Vv. oesophageales [eso—]
V. vertebralis
 V. occipitalis
 V. vertebralis anterior[105]
 (V. vertebralis accessoria)[105]
Plexus venosus suboccipitalis
V. cervicalis profunda
Vv. thoracicae internae
 Vv. epigastricae superiores
 Vv. subcutaneae abdominis
 Vv. musculophrenicae
 Vv. intercostales anteriores
V. intercostalis suprema
V. intercostalis superior sinistra

VENA JUGULARIS INTERNA[106]

Bulbus superior venae jugularis
V. aquēductus cochleae
Bulbus inferior venae jugularis
Plexus pharyngeus [pharyngealis]
Vv. pharyngeales
Vv. meningeae
V. lingualis
 Vv. dorsales linguae
 V. comitans nervi hypoglossi
  V. sublingualis
 V. profunda linguae
V. thyroidea superior
Vv. thyroideae mediae
V. sternocleidomastoidea
V. laryngea superior

Vena facialis
V. angularis
Vv. supratrochleares
V. supraorbitalis
Vv. palpebrales superiores
Vv. nasales externae
Vv. palpebrales inferiores
V. labialis superior
Vv. labiales inferiores

---

[105] A small vein accompanying the *Arteria cervicalis ascendens*. It begins as a venous plexus adjacent to the more cranial transverse processes and ends in the *Vena vertebralis*. The *Vena vertebralis accessoria* is inconstant; it drains the vertebral venous plexus and descends through the seventh cervical foramen transversarium.

[106] The modes of termination of the tributaries of the *Vena jugularis interna* are most variable. The arrangement used here is merely representative.

V. profunda faciei [facialis]
Rami parotidei
V. palatina externa
V. submentalis

Vena retromandibularis
Vv. temporales superficiales
V. temporalis media
V. transversa faciei [facialis]
Vv. maxillares
Plexus pterygoideus
    Vv. meningeae mediae
    Vv. temporales profundae
    V. canalis pterygoidei
    Vv. auriculares anteriores
    Vv. parotideae
    Vv. articulares
    Vv. tympanicae
    V. stylomastoidea

VENA JUGULARIS EXTERNA
V. auricularis posterior
V. jugularis anterior
    Arcus venosus jugularis
V. suprascapularis
Vv. transversae cervicis

SINUS DURAE MATRIS
Sinus transversus
Confluens sinuum
Sinus occipitalis
Plexus basilaris
Sinus sigmoideus
Sinus sagittalis superior
    Lacunae laterales
Sinus sagittalis inferior
Sinus rectus
Sinus petrosus inferior
    Vv. labyrinthi
Sinus petrosus superior
Sinus cavernosus
    Sinus intercavernosi
Sinus sphenoparietalis

Venae diploicae
V. diploica frontalis
V. diploica temporalis anterior
V. diploica temporalis posterior
V. diploica occipitalis

Venae emissariae[107]
V. emissaria parietalis
V. emissaria mastoidea
V. emissaria condylaris
V. emissaria occipitalis
Plexus venosus canalis hypoglossi
Plexus venosus foraminis ovalis
Plexus venosus caroticus internus

VENAE CEREBRI[108]

Venae superficiales cerebri
    Vv. superiores cerebri
        Vv. prefrontales
        Vv. frontales
        Vv. parietales
        Vv. occipitales
    V. mediae superficiales cerebri
        V. anastomotica inferior
        V. anastomotica superior
    Vv. inferiores cerebri
        V. unci

Venae profundae cerebri
    V. basalis
        Vv. anteriores cerebri
        V. media profunda cerebri
            Vv. insulares
        Vv. thalamostriatae inferiores
        V. gyri olfactorii
        V. ventricularis inferior
        V. choroidea inferior
        Vv. pedunculares
    V. magna cerebri
        Vv. internae cerebri
            V. choroidea superior
        V. thalamostriata superior [V. terminalis]
        V. anterior septi pellucidi
        V. posterior septi pellucidi
        V. medialis atrii (ventriculi lateralis)
        V. lateralis atrii (ventriculi lateralis)
        Vv. nuclei caudati
        Vv. directae laterales
        V. posterior corporis callosi
        V. dorsalis corporis callosi

Venae trunci encephalici
    V. pontomesencephalica anterior

---

[107] Only the more usual emissary veins are included.
[108] The list of cerebral veins has been amplified. See footnote 96.

Vv. pontis
Vv. medullae oblongatae
V. recessus lateralis ventriculi quarti

Venae cerebelli
V. superior vermis
V. inferior vermis
Vv. superiores hemispherii cerebelli
Vv. inferiores hemispherii cerebelli
V. precentralis cerebelli
V. petrosa

Vena ophthalmica superior
V. nasofrontalis
Vv. ethmoidales
V. lacrimalis
Vv. vorticosae [Vv. choroideae oculi]
Vv. ciliares
Vv. ciliares anteriores
   (Sinus venosus sclerae)
   Vv. sclerales
V. centralis retinae
Vv. episclerales
   Vv. palpebrales
   Vv. conjunctivales

Vena ophthalmica inferior

VENA SUBCLAVIA
Vv. pectorales
V. scapularis dorsalis
(V. thoracoacromialis)

Vena axillaris
V. thoracica lateralis
Vv. thoracoepigastricae
Plexus venosus areolaris

VENAE SUPERFICIALES MEMBRI SUPERIORIS
   V. cephalica
      V. thoracoacromialis
      (V. cephalica accessoria)
   V. basilica
   V. intermedia cubiti
   V. intermedia antebrachii
   V. intermedia cephalica
   V. intermedia basilica
   Rete venosum dorsale manus

Vv. intercapitulares
Arcus venosus palmaris superficialis
   Vv. digitales palmares
   Vv. metacarpales dorsales

VENAE PROFUNDAE MEMBRI SUPERIORIS
   Vv. brachiales
   Vv. ulnares
   Vv. radiales
   Arcus venosus palmaris profundus
      Vv. metacarpales palmares

VENA AZYGOS

Arcus venae azygou
V. intercostalis superior dextra
V. hemiazygos
V. hemiazygos accessoria
V. intercostalis superior sinistra
Vv. oesophageales [eso—]
Vv. bronchiales
Vv. pericardiales
Vv. mediastinales
Vv. phrenicae superiores
V. lumbalis ascendens
   Vv. lumbales
V. subcostalis
Vv. intercostales posteriores[109]
   Ramus dorsalis
   V. intervertebralis
   Ramus spinalis

VENAE COLUMNAE VERTEBRALIS
Plexus venosus vertebralis externus anterior
Plexus venosus vertebralis externus posterior
Plexus venosus vertebralis internus anterior
   Vv. basivertebrales
   Vv. spinales anteriores/posteriores
Plexus venosus vertebralis internus posterior

VENA CAVA INFERIOR

Vv. phrenicae inferiores[110]
Vv. lumbales
V. lumbalis ascendens
Venae hepaticae
   Vv. hepaticae dextrae

---

[109] The 4th to 11th *Venae intercostales posteriores* are tributaries of the *Vena azygos* on the right and of the *vena hemiazygos* on the left.
[110] So-named to distinguish them from the *Venae phrenicae superiores.*

Vv. hepaticae intermediae
Vv. hepaticae sinistrae
Venae renales
  V. suprarenalis sinistra
  V. testicularis sinistra
  V. ovarica sinistra
V. suprarenalis dextra
V. testicularis dextra
V. ovarica dextra
Plexus pampiniformis

VENA PORTAE HEPATIS[111]
Ramus dexter
  Ramus anterior
  Ramus posterior
Ramus sinister
  Pars transversa
    Rami caudati
  Pars umbilicalis[112]
    Ligamentum venosum
    (*Ductus venosus*)
    Rami laterales
    V. umbilicalis sinistra
    Ligamentum teres hepatis
    Rami mediales
V. cystica
Vv. paraumbilicales
V. gastrica sinistra
V. gastrica dextra
V. prepylorica

VENA MESENTERICA SUPERIOR
Vv. jejunales
Vv. ileales
V. gastro-omentalis [epiploica] dextra
Vv. pancreaticae
Vv. pancreaticoduodenales
V. ileocolica
  V. appendicularis
V. colica dextra
V. colica media [intermedia]

VENA SPLENICA
Vv. pancreaticae
Vv. gastricae breves
V. gastro-omentalis [epiploica] sinistra
V. mesenterica inferior
  V. colica sinistra

Vv. sigmoideae
V. rectalis superior

VENA ILIACA COMMUNIS
V. sacralis mediana
V. iliolumbalis

VENA ILIACA INTERNA
Vv. gluteae superiores
Vv. gluteae inferiores
Vv. obturatoriae
Vv. sacrales laterales
Plexus venosus sacralis
Plexus venosus rectalis
Vv. vesicales
Plexus venosus vesicalis
Plexus venosus prostaticus
V. dorsalis profunda penis
V. dorsalis profunda clitoridis
Vv. uterinae
Plexus venosus uterinus
Plexus venosus vaginalis
Vv. rectales mediae
V. pudenda interna
  Vv. profundae penis
  Vv. profundae clitoridis
  Vv. rectales inferiores
Vv. scrotales posteriores
Vv. labiales posteriores
V. bulbi penis
V. bulbi vestibuli

VENA ILIACA EXTERNA
V. epigastrica inferior
V. circumflexa iliaca profunda

VENAE SUPERFICIALES MEMBRI INFERIORIS
Vena saphena magna
  Vv. pudendae externae
  V. circumflexa superficialis ilium
  V. epigastrica superficialis
  V. saphena accessoria
  Vv. dorsales superficiales penis
  Vv. dorsales superficiales clitoridis
  Vv. scrotales anteriores
  Vv. labiales anteriores
  V. saphena parva
    Rete venosum dorsale pedis

---

[111] Only the most widely accepted branches of the *vena portae hepatis* have been included.
[112] The name *Sinus* can be applied to this part, always dilated.

Arcus venosus dorsalis pedis
Vv. metatarsales dorsales pedis
Vv. digitales dorsales pedis
Rete venosum plantare
Arcus venosus plantaris
Vv. metatarsales plantares
Vv. digitales plantares
Vv. intercapitulares
V. marginalis lateralis
V. marginalis medialis

VENAE PROFUNDAE MEMBRI INFERIORIS

Vena femoralis[113]
V. profunda femoris
Vv. circumflexae mediales femoris
Vv. circumflexae laterales femoris
Vv. perforantes

Vena poplitea
Vv. geniculares
Vv. tibiales anteriores
Vv. tibiales posteriores
Vv. fibulares [peroneales]

SYSTEMA LYMPHATICUM

VASA LYMPHATICA

Vas lymphocapillare
Rete lymphocapillare[114]
Vas lymphaticum
Plexus lymphaticus
Vas lymphaticum superficiale
Vas lymphaticum profundum

TRUNCI LYMPHATICI

Truncus lumbaris dexter/sinister
Trunci intestinales
Truncus bronchomediastinalis dexter/-sinister
Truncus subclavius dexter/sinister
Truncus jugularis dexter/sinister

DUCTUS LYMPHATICI

Ductus lymphaticus dexter (Ductus thoracicus dexter)[115]
Ductus thoracicus
Arcus ductus thoracici
Pars cervicalis
Pars thoracica
Pars abdominalis
Cisterna chyli

NODI REGIONALES[116]

(CAPUT ET COLLUM)
Nodi lymphatici[117]
occipitales
mastoidei
parotidei superficiales
parotidei profundi
preauriculares
infra-auriculares
intraglandulares
faciales
(Nodus buccinatorius
nasolabialis
malaris
mandibularis)
submentales
submandibulares
cervicales anteriores
superficiales
profundi
prelaryngeales
thyroidei
pretracheales
paratracheales
cervicales laterales
superficiales
profundi
jugulares laterales
jugulares anteriores
Nodus jugulodigastricus
Nodus jugulo-omohyoideus
supraclaviculares
retropharyngeales

[113] The tributaries of the *V. femoralis* often differ from the arrangement shown here.
[114] These are two terms transferred from *Nomina Histologica.*
[115] Typically formed by the right jugular, subclavian, and bronchomediastinal lymphatic trunks, any one of which may, however, end separately in the right brachiocephalic vein.
[116] A number of regional groups of lymphnodes have been added to this list.
[117] Many additional groups of regional lymph nodes have been included. Those which are somewhat variable are enclosed in parentheses, so ( ).

(Membrum superius)
Plexus lymphaticus axillaris
Nodi lymphatici axillares
  cubitales
  superficiales
  profundi
  brachiales
  interpectoralles

(Thorax)
Nodi lymphatici
  paramammarii
  parasternales
  intercostales
  prevertebrales
  phrenici superiores
  prepericardiales
  pericardiales laterales
  mediastinales anteriores
  (Nodus ligamentis arteriosi)
  mediastinales posteriores
    juxta-esophageales pulmonales
    tracheobronchiales
      superiores
      inferiores
    paratracheales
    (Nodus arcus venae azygos)

(Abdomen—Nodi parietales)
Nodi lymphatici
  lumbales [lumbares] sinistri
    aortici laterales
    pre-aortici
    postaortici
  lumbales [lumbares] intermedii
  lumbales [lumbares] dextri
    cavales laterales
    precavales
    postcavales
  phrenici inferiores
  epigastrici inferiores

(Abdomen–Nodi viscerales)
Nodi lymphatici
  coeliaci
  gastrici (dextri/sinistri)
  (Annulus [Anulus] lymphaticus cardiae)
  gastro-omentales (dextri/sinistri)
  pylorici
    (Nodus suprapyloricus
    Nodi subpylorici
    Nodi retropylorici)

pancreatici
  superiores
  inferiores
splenici [lienales]
pancreaticoduodenales
  superiores
  inferiores
hepatici
  Nodus cysticus
  Nodus foraminalis
mesenterici
  juxta-intestinales
  superiores (centrales)
ileocolici
precaecales [-cecales]
retrocaecales [-cecales]
appendiculares
mesocolici
  paracolici
  colici (dextri/medii/sinistri)
mesenterici inferiores
  sigmoidei
  rectales superiores

(Pelvis—Nodi parietales)
Nodi lymphatici
  iliaci communes
    mediales
    intermedii
    laterales
    subaortici
    promontorii
  iliaci externi
    mediales
    intermedii
    laterales
    (Nodus lacunaris medialis
    Nodus lacunaris intermedius
    Nodus lacunaris lateralis)
    interiliaci
    obturatorii
  iliaci interni
    gluteales
      superiores
      inferiores
    sacrales

(Pelvis—Nodi viscerales)
Nodi lymphatici
  paravesiculares
    prevesiculares
    postvesiculares

vesicales laterales
para-uterini
paravaginales
pararectales [anorectales]

(MEMBRUM INFERIUS)
Nodi lymphatici
  inguinales
    superficiales
      superomediales
      superolaterales
      inferiores
    profundi
  popliteales
    superficiales
    profundi
  (Nodus tibialis anterior
  Nodus tibialis posterior
  Nodus fibularis)

SPLEN [LIEN]

Splen accessorius
Facies diaphragmatica
Facies visceralis
  Facies renalis
  Facies gastrica
  Facies colica
Extremitas anterior
Extremitas posterior
Margo inferior
Margo superior
Hilum splenicum
Tunica serosa
Tunica fibrosa
Trabeculae splenicae
Pulpa splenica
Sinus splenicus
Rami splenici (lienales)
Penicilli
Folliculi lymphatici splenici [Lymphonoduli
  splenici]

SYSTEMA NERVOSUM

Neurona, Neurofibrae, Terminationes
nervorum, Synapses and Neurologia—
see Nomina Histologica
Neurogenesis—*see Nomina Embryologica*

MENINGES

DURA MATER ENCEPHALI
Falx cerebri
Tentorium cerebelli
  Incisura tentorii
Falx cerebelli
Diaphragma sellae
Cavum trigeminale
Spatium subdurale

DURA MATER SPINALIS
Filum terminale externum [durale]
Cavitas [Cavum] epiduralis

ARACHNOIDEA MATER ENCEPHALI
Cavitas [Cavum] subarachnoidea
  Liquor cerebrospinalis

Cisternae subarachnoideae
  Cisterna cerebellomedullaris
  Cisterna fossae lateralis cerebri
  Cisterna chiasmatis
  Cisterna interpeduncularis
  Granulationes arachnoideales

ARACHNOIDEA MATER SPINALIS
Cavitas [Cavum] subarachnoidea
  Liquor cerebrospinalis

PIA MATER ENCEPHALI
Tela choroidea ventriculi quarti
Plexus choroideus ventriculi quarti
Tela choroidea ventriculi tertii
Plexus choroideus ventriculi tertii
Plexus choroideus ventriculi lateralis
Glomus choroideum

PIA MATER SPINALIS
Ligamentum denticulatum
Septum cervicale intermedium
Filum terminal internum [pialis]

PARS CENTRALIS
  [Systema Nervosum Centrale]

Substantia grisea (Nuclei et Columnae)
Substantia alba (Tractus et Fasciculi)
Formatio [Substantia] reticularis
Substantia gelatinosa
Ependyma

MEDULLA SPINALIS

Intumescentia cervicalis
Intumescentia lumbosacralis
Conus medullaris
Filum terminale (spinale)
Ventriculus terminalis
Fissura mediana ventralis [anterior]
Sulcus medianus dorsalis [posterior]
   Septum medianum dorsale [posterius]
Sulcus ventrolateralis [anterolateralis]
Sulcus dorsolateralis [posterolateralis]
Sulcus intermedius dorsalis [posterior]
Funiculi medullae spinalis
   Funiculus ventralis [anterior]
   Funiculus lateralis
   Funiculus dorsalis [posterior]
Segmenta medullae spinalis[118]
   Cervicalia (1–8) = Pars cervicalis
   Thoracica (1–12) = Pars thoracica
   Lumbalia (1–5) = Pars lumbalis
   Sacralia (1–5) = Pars sacralis
   Coccygea (1–3) = Pars coccygea

SECTIONES MEDULLAE SPINALIS
Canalis centralis
Substantia grisea
Substantia alba
Substantia gelatinosa centralis

COLUMNAE GRISEAE

Columna ventralis [anterior][119]
   Cornu ventrale [anterius][120]
     Nucleus ventrolateralis
     Nucleus ventromedialis
     Nucleus dorsolateralis
     Nucleus retrodorsolateralis[121]

Nucleus dorsomedialis
Nucleus centralis
Nucleus nervi accessorii [Nuc.
   accessorius]
Nucleus nervi phrenici [Nuc.
   phrenicus]

Columna dorsalis [posterior]
   Cornu dorsale [posterius]
     Apex cornus dorsalis [posterioris]
     Caput cornus dorsalis [posterioris]
     Cervix cornus dorsalis [posterioris]
     Basis cornus dorsalis [posterioris]
   Substantia gelatinosa
   Substantia visceralis secundaria[121]

Columna lateralis
   Cornu laterale
Substantia (grisea) intermedia centralis
   Columna thoracica [Nuc. thoracicus][122]
Substantia (grisea) intermedia lateralis
   Columna intermediolateralis
     [autonomica][123]
   Nuclei parasympathici sacrales[124]
Formatio reticularis

SUBSTANTIA ALBA
   Commissura alba

Funiculus ventralis [anterior][125]
Fasciculi propii ventrales [anteriores]
   Fasciculus sulcomarginalis
Tractus corticospinalis [pyramidalis]
   ventralis [anterior]
Tractus vestibulospinalis[126]
Tractus reticulospinalis[127]
Tractus spinothalamicus ventralis [anterior]

---

[118] Numbers are noted with each segmental region. This is not to teach the obvious but to suggest a convenient (if unscholarly) international abbreviation. Thus *segmentum cervicale 5* may serve *in print* for *segm. cerv. quintum* in speech.

[119] The *columnae griseae* and *cornua* are really synonyms, but the latter term is usually reserved for appearances in transverse sections of the spinal cord. Although *cornu* is much used, *columna* is perhaps preferable on *all* occasions.

[120] The *nucleus retrodorsolateralis* consists of neurons considered to innervate the digital muscles in primates; it occurs in C8, T1 and S1 to S3.

[121] *Substanti visceralis secundaria* is dorsal to *substantia intermedia centralis*.

[122] Usual extent is segm. T1 to L1 or 2.

[123] Usual extent is segm. C7 to L2.

[124] Usually limited to segm. S2 to 4.

[125] The term *funiculus* has been reserved for the "white" columns. *Tractus* is the preferable term for bundles of nerve fibers of like connexions, but *fasciculus* is still customary for certain tracts.

[126] More than one such tract is considered to exist in some species, including mankind.

[127] The status of the reticulospinal tracts in mankind is still *sub judice*.

Funiculus lateralis
Fasciculi proprii laterales
Tractus corticospinalis [pyramidalis]
  lateralis
Tractus rubrospinalis
Tractus bulboreticulospinalis[127]
Tractus pontoreticulospinalis[127]
Tractus tectospinalis
Tractus olivospinalis
Tractus spinotectalis
Tractus spinothalamicus lateralis
Tractus spinocerebellaris ventralis [anterior]
Tractus spinocerebellaris dorsalis [posterior]
Tractus dorsolateralis
Tractus spino-olivaris
Tractus spinoreticularis

Funiculus dorsalis [posterior]
Fasciculi proprii dorsales [posteriores]
  Fasciculus septomarginalis
  Fasciculus interfascicularis [semilunaris]
Fasciculus gracilis
Fasciculus cuneatus

ENCEPHALON

TRUNCUS ENCEPHALI[128]
RHOMBENCEPHALON

MEDULLA OBLONGATA [BULBUS, MYELENCEPHALON]

Fissura mediana ventralis [anterior]
Pyramis (medullae oblongatae)
Decussatio pyramidum [Dec. motoria]
Sulcus ventrolateralis [anterolateralis]
Funiculus lateralis
Oliva
Fibrae arcuatae externae ventrales
  [anteriores]
Sulcus retro-olivaris
Area retro-olivaris
Sulcus dorsolateralis [posterolateralis]

Pedunculus cerebellaris caudalis [inferior]
Tuberculum trigeminale[129]
Fasciculus cuneatus
Tuberculum cuneatum
Fasciculus gracilis
Tuberculum gracile
Sulcus medianus dorsalis [posterior]

SECTIONES MEDULLAE OBLONGATAE
Fasciculus pyramidalis[130]
  Fibrae corticospinales
  Fibrae corticonucleares
Decussatio pyramidum [Dec. motoria]
Fasciculus gracilis
Nucleus gracilis
Fasciculus cuneatus
Nucleus cuneatus
Nucleus cuneatus accessorius
Fibrae arcuatae internae
Decussatio lemniscorum medialium [Dec.
  sensoria]
Lemniscus medialis
Tractus tectospinalis
Fasciculus longitudinalis medialis
Fasciculus longitudinalis dorsalis
Tractus spinalis nervi trigemini
Nucleus spinalis [inferior] nervi trigemini
Formatio [Substantia] reticularis
Nucleus olivaris caudalis [inferior]
  Amiculum olivare
  Hilum nuclei olivaris caudalis [inferioris]
Nucleus olivaris accessorius medialis
Nucleus olivaris accessorius dorsalis
  [posterior]
Tractus spino-olivaris
Tractus olivocerebellaris
Pedunculus cerebellaris caudalis [inferior]
Nucleus nervi hypoglossi [Nucleus
  hypoglossalis]
Nucleus paramedianus dorsalis [posterior]
Nucleus dorsalis nervi vagi [Nucleus vagalis
  dorsalis][131]
Nucleus intercalatus
Tractus solitarius
Nucleus solitarius

[128] A new term to express the concept "brain stem," including *myelencephalon (med. obl.), metencephalon (pons)*, and *mesencephalon. Rhombencephalon = myelencephalon + metencephalon.*
[129] Corresponds to descending tract or *radix* spinalis of the trigeminal nerve.
[130] *Fasciculus* preferred because of the mixed nature of this "tract."
[131] The cranial pole of this dorsal vagal nucleus is sometimes identifed as the *nucleus dorsalis nervi glossopharyngei.*

A 65

Nucleus parasolitarius
Nuclei vestibulares[132]
Nucleus vestibularis caudalis [inferior]
Nucleus vestibularis medialis
Nucleus vestibularis lateralis
Nuclei cochleares[132]
   Nucleus cochlearis ventralis [anterior]
   Nucleus cochlearis dorsalis [posterior]
Nucleus commissuralis
Nucleus ambiguus
Nucleus salivatorius caudalis [inferior][131]
Nuclei arcuati
Fibrae arcuatae externae ventrales
   [anteriores]
Fibrae arcuatae externae dorsales
   [posteriores]
Raphe medullae oblongatae
   Nuclei raphae

PONS [METENCEPHALON]

Sulcus bulbopontinus
Sulcus basilaris
Pedunculus cerebellaris medius [pontinus]
Trigonum pontocerebellare

SECTIONES PONTIS

Pars ventralis [basilaris] pontis
Fibrae pontis longitudinales
   Fibrae corticospinales
   Fibrae corticonucleares
   Fibrae corticoreticulares
   Fibrae corticopontinae
Fibrae pontis transversae
   Fibrae pontocerebellares
Nuclei pontis

Pars dorsalis pontis [Tegmentum pontis]
Raphe pontis
Fasciculus longitudinalis medialis
Fasciculus longitudinalis dorsalis
Lemniscus medialis
Tractus tectospinalis

Formatio reticularis
Lemniscus spinalis
Tractus spinalis nervi trigemini
Nucleus spinalis [inferior] nervi trigemini
Nucleus pontinus nervi trigemini[133]
Lemniscus trigeminalis [Tractus
   trigeminothalamicus]
Tractus mesencephalicus nervi trigemini
   (Tr. mes. trigeminalis)
Nucleus mesencephalicus nervi trigemini
   (Nuc. mes. trigeminalis)[134]
Nucleus motorius nervi trigemini (Nuc.
   mot. trigeminalis)
Nucleus nervi abducentis (Nuc. abducens)
Nucleus nervi facialis (Nuc. facialis)
Genu nervi facialis
Nucleus salivatorius rostralis [superior][135]
Nucleus lacrimalis
Nucleus olivaris rostralis [superioris][135]
   Tractus olivocochlearis
Nuclei vestibulares (*see* Medulla oblongata)
   Nucleus vestibularis medialis
   Nucleus vestibularis lateralis
   Nucleus vestibularis rostralis [superior][135]
Nuclei cochleares (*see* Medulla oblongata)
Corpus trapezoideum
Nucleus ventralis corporis trapezoidei
Nucleus dorsalis corporis trapezoidei
Lemniscus lateralis
Nuclei lemnisci lateralis

VENTRICULUS QUARTUS

Fossa rhomboidea
Recessus lateralis (ventriculi quarti)
Sulcus medianus
Eminentia medialis
Colliculus facialis
Sulcus limitans
Area vestibularis
Fovea rostralis [superior]
Locus coeruleus
Fovea caudalis [inferior]
Striae medullares (ventriculi quarti)

---

[132] Certain of these nuclei occur at pontine and medullary levels and are hence mentioned in both places.
[133] Many variant names for this nucleus, each with some advantages, have been used or suggested. Its sensory nature is well known, and the term "principal" is perhaps misleading. An adjective denoting *position* was therefore preferred by the Subcommittee.
[134] The term *tractus* has been omitted from this term as an unnecessary complication. It still remains too lengthy and could be improved to *tr. mesencephalicus trigeminalis*.
[135] When at the level or inside the skull—*rostralis*; when not—*cranialis*.

Trigonum nervi hypoglossi [Tri. hypoglossale]
Funiculus separans
Trigonum nervi vagi [Tri. vagale]
Area postrema
Tegmen ventriculi quarti
 Velum medullare rostralis [superius] [anterius]
  Frenulum veli medullaris rostralis [superius]
 Velum medullare caudale [inferius] [posterius]
Tela choroidea ventriculi quarti
Plexus choroideus ventriculi quarti
Taenia [Tenia] ventriculi quarti
 Obex
Apertura mediana ventriculi quarti
Apertura lateralis ventriculi quarti

CEREBELLUM

Folia cerebelli
Fissure cerebelli
Vallecula cerebelli

CORPUS CEREBELLI
 Vermis cerebelli
 Hemispherium cerebelli

Lobus rostralis [anterior] cerebelli
 Lingula
 Lobulus centralis
 Culmen
 Ala lobuli centralis
 Lobulus quadrangularis [Pars rostralis (anterior)]
Fissura prima

Lobus caudalis [posterior] cerebelli
 Declive
 Folium vermis
 Tuber vermis
 Pyramis vermis
  Fissura secunda[136]
 Uvula vermis

Lobulus simplex [Lobulus quadrangularis [Pars caudalis/posterior]
Lobulus semilunaris rostralis [superior]
 Fissura horizontalis
Lobulus semilunaris caudalis [inferior][137]
Lobulus gracilis [Lobulus paramedianus]
Lobulus biventer
Tonsilla cerebelli
Fissura dorsolateralis [posterolateralis]

Lobus flocculonodularis
 Nodulus
 Flocculus
  Pedunculus flocculi
 Paraflocculus

Archaeocerebellum [Archeo-]
Palaeocerebellum [Palaeo]
Neocerebellum

SECTIONES CEREBELLI
Arbor vitae cerebelli
Corpus medullare
 Laminae albae
Cortex cerebelli
 Stratum moleculare [plexiforme]
 Stratum neuronorum piriformium[138]
 Stratum granulosum
Nuclei cerebelli
 Nucleus dentatus
  Hilum nuclei dentati
 Nucleus emboliformis
 Nucleus globosus
 Nucleus fastigii [fastigiatus]
Pedunculi cerebelli
 Pedunculus cerebellaris caudalis [inferior]
 Pedunculus cerebellaris medius [pontinus]
 Pedunculus cerebellaris rostralis [superior]

MESENCEPHALON

PEDUNCULUS CEREBRI [CEREBRALIS]
 Pars ventralis [anterior] [Crus cerebri]
 Pars dorsalis [posterior][139]

---

[136] This separates *pyramis* and *uvula* in the vermis cerebelli.
[137] This was *lobulus simplex* in *Nomina Anatomica*, third edition.
[138] This new term identifies the "layer" of Purkinje neurons.
[139] *Tegmentum = partes dorsales* of the *pedunculi cerebri*. On each side it extends from the *subst. nigra* to the level of the *aqueductus mes.*

A 67

Trigonum lemnisci
Pedunculus cerebellaris rostralis [superior]
  Fossa interpeduncularis
  Substantia perforata interpeduncularis
    [posterior]
TECTUM MESENCEPHALI
  Lamina tecti [tectalis]
  Colliculus caudalis [inferior]
  Colliculus rostralis [superior]
  Brachium colliculi caudalis [inferioris]
  Brachium colliculi rostralis [superioris]

AQUEDUCTUS MESENCEPHALI [CEREBRI]

SECTIONES MESENCEOPHALI

BASIS PEDUNCULI CEREBRI
  Fibrae corticospinales[140]
  Fibrae corticonucleares[140]
  Fibrae corticopontinae
  Fibrae parietotemporopontinae
  Fibrae frontopontinae

SUBSTANTIA NIGRA
  Pars compacta
  Pars reticularis

TEGMENTUM MESENCEPHALI[139]
  Substantia grisea centralis
  Formatio reticularis
  Fibrae corticoreticulares
  Fasciculus longitudinalis medialis
  Fasciculus longitudinalis dorsalis
  Tractus mesencephalicus nervi trigemini
    [trigeminalis]
  Nucleus tractus mesencephalici nervi
    trigemini [N. mesencephalicus
    trigeminalis]
  Nucleus nervi oculomotorii [Nuc.
    oculomotorius]
    Nucleus oculomotorius accessorius
    (autonomicus)
  Nucleus nervi trochlearis [Nuc.
    trochlearis]
  Nucleus interpeduncularis
  Nucleus interstitialis
  Nuclei tegmenti [tegmentales]

Nucleus ruber
  Pars magnocellularis
  Pars parvocellularis
Tractus tegmentalis centralis
Decussationes tegmenti[141]
Decussatio pedunculorum cerebellarium
  rostralium [superiorum]
Fibrae dentato rubrales
Tractus rubrospinalis
Tractus tectobulbaris
Tractus tectospinalis
Lemniscus lateralis
Lemniscus medialis
Lemniscus spinalis
Lemniscus trigeminalis

TECTUM MESENCEPHALI
Lamina tecti
Nucleus colliculi caudalis [inferioris]
Brachium colliculi caudalis [inferioris]
Commissura colliculorum caudalium
  [inferiorum]
Strata (grisea et alba) colliculi rostralis
  [superioris]
Brachium colliculi rostralis [superioris]
Commissura colliculorum rostralium
  [superiorum]
Decussatio trochlearis [D. nervorum
  trochlearium]

PROSENCEPHALON

DIENCEPHALON

EPITHALAMUS
Habenula
Sulcus habenulae [habenularis]
Trigonum habenulae [habenularis]
Commissura habenularum [habenularis]
Commissura epithalamica
Corpus pineale [glandula pinealis]

SECTIONES EPITHALAMI
Nuclei habenulae medialis et lateralis
Tractus habenulo-interpeduncularis

[140] Collectively these fibres have been commonly known as the *fasciculus (tractus) pyramidalis*. Since the term tractus implies a uniformity of connexions and functions now known to be incorrect, the new terms are preferable.
[141] These include decussations of the rubrospinal and rubroreticular tracts (ventral) and of the tectospinal tracts (dorsal).

Commissura habenularum [habenularis]
Area pretectalis
Nuclei pretectales
Commissura epithalamica [posterior]
Corpus pineale [glandula pinealis]
Organum subcommissurale

THALAMUS DORSALIS[142]
Adhesio interthalamica
Tuberculum anterius thalami
Stria medullaris thalami
Pulvinar

METATHALAMUS
Corpus geniculatum mediale
Corpus geniculatum laterale

THALAMUS VENTRALIS[142]

HYPOTHALAMUS
Area preoptica
Chiasma opticum
Tractus opticus
  Radix lateralis
  Radix medialis
Corpus mamillare
Tuber cinereum
Infundibulum
Neurohypophysis (*see* page A 46, 47)

VENTRICULUS TERTIUS
Sulcus hypothalamicus
Foramen interventriculare
Recessus opticus
Recessus infundibuli [infundibularis]
Recessus pinealis
Recessus suprapinealis
Tela choroidea ventriculi tertii
Taenia [tenia] thalami
Plexus choroideus ventriculi tertii
Organum subfornicale

SECTIONES THALAMI ET METATHALAMI[143]
Laminae medullares thalami

Nuclei reticulares (thalami)
Nuclei anteriores (thalami)
  Nucleus anterodorsalis
  Nucleus anteroventralis
  Nucleus anteromedialis
Nuclei mediani (thalami)
  Nuclei paraventriculares anteriores et
    posteriores
  Nucleus rhomboidalis
  Nucleus reuniens
Nuclei mediales (thalami)
  Nucleus medialis dorsalis
Laminae medullares interna et externa
Nuclei intralaminares (thalami)
  Nucleus centromedianus
  Nucleus paracentralis
  Nucleus parafascicularis
  Nucleus centralis lateralis
  Nucleus centralis medialis
Nuclei ventrolaterales (thalami)
  Nucleus lateralis posterior
  Nucleus lateralis dorsalis
  Nucleus ventralis anterior
  Nucleus ventralis lateralis
  Nucleus ventralis medialis
  Nuclei ventrales posteriores
    Nucleus ventralis posterolateralis
    Nucleus ventralis posteromedialis
Nuclei posteriores (thalami)
  Nuclei pulvinares
  Nucleus (corporis geniculati) lateralis
    (Pars dorsalis)[144]
  Nucleus (corporis geniculati) medialis
    (Pars dorsalis)
Nucleus corporis geniculati lateralis
  (Pars ventralis)
Nucleus corporis geniculati medialis
  (Pars ventralis)
Nucleus subthalamicus
Nuclei reticulares (thalami)
Zona incerta
  Nuclei areae H, H$_1$, H$_2$[145]

Tractus et fasciculi thalamici
Lemniscus lateralis

---

[142] These expressions have wide currency in research and must now be included.
[143] It is probably impossible to draw up a completely satisfactory nominal list of thalamic nuclei. Doubtless the nomenclature set out here will require amendments as disagreements between the experts are solved.
[144] The more widely used *nucleus geniculatus* is a desirable simplification in these lengthy names.
[145] These are the areas of Forel.
[145] The *Nucleus entopeduncularis*, relatively small in man is located in the internal capsule adjacent to the medial edge of the *Globus pallidus*, dorsolateral to the *Nucleus hypothalamicus lateralis*.

Lemniscus medialis
Lemniscus spinalis
Lemniscus trigeminalis
Brachium colliculi caudalis [inferioris]
Radiatio acustica
Brachium colliculi rostralis [superioris]
Radiatio optica
Radiationes thalamicae anteriores
Radiationes thalamicae centrales
Radiationes thalamicae posteriores
Tractus dentatothalamicus
Fasciculus thalamicus
Fasciculus subthalamicus
Fasciculus mamillothalamicus
Pedunculus thalami caudalis [inferior]
Ansa et fasciculus lenticulares
Ansa et fasciculus pedunculares
Fibrae intrathalamicae
Fibrae periventriculares

SECTIONES HYPOTHALAMI
Regio [area] hypothalamica dorsalis
    Nucleus entopeduncularis[146]
    Nucleus ansae lenticularis
    Regio hypothalamica anterior[147]
    Nuclei preoptici mediales et laterales
    Nucleus supraopticus
    Nuclei paraventriculares
    Nucleus hypothalamicus anterior
Regio hypothalamica intermedia
    Nuclei tuberales
    Area hypothalamica lateralis
    Nucleus hypothalamicus ventromedialis
    Nucleus hypothalamicus dorsomedialis
    Nucleus hypothalamicus dorsalis
    Nucleus periventricularis posterior
    Nucleus infundibularis [arcuatus]
Regio hypothalamica posterior
    Nuclei corporis mamillaris mediales et
        laterales
    Nucleus hypothalamicus posterior

Neurohypophysis
Tractus et fasciculi hypothalamici
    Fibrae periventriculares
    Commissura supraoptica dorsalis[148]
    Commissura supraoptica ventralis
    Fasciculus longitudinalis dorsalis
    Fasciculus mamillotegmentalis
    Fasciculus mamillothalamicus

Fornix
Fibrae striae terminalis
Fasciculus prosencephalicus medialis[149]
Tractus hypothalamohypophysialis[150]
    Fibrae supraopticae
    Fibrae paraventriculares
Tractus supraopticohypophysialis
Tractus paraventriculohypophysialis[150]

TELENCEPHALON

CEREBRUM

Cortex cerebri [Pallium]
Gyri cerebri
Sulci cerebi
Lobi cerebri
Fissura longitudinalis cerebri
Fissura transversa cerebri
Fossa lateralis cerebri
Margo superior (superomedialis)
Margo inferior (inferolateralis)
Margo medialis (inferomedialis)

HEMISPHERIUM CEREBRI [CEREBRALIS]

FACIES SUPEROLATERALIS (HEMISPHERII)
Sulcus centralis
Sulcus lateralis

---

146   The *Nucleus entopeduncularis*, relatively small in man is located in the internal capsule adjacent to
       the medial edge of the *Globus pallidus*, dorsolateral to the *Nucleus hypothalamicus lateralis*.
147   The *Nucleus ansae lenticularis* is located within the *Ansae lenticularis* as it curves round the medial
       edge of the *Globus pallidus*.
148   These commissures, associated with the names of van Gudden, Meynert and Ganser, are probably in
       fact decussations. At least three supraoptic "commissures" have been described, the most central being
       apparently absent in primates. Details of their connexions in the human brain are uncertain.
149   A new term for the "medial forebrain bundle," the main pathway for longitudinal connexions in the
       hypothalamus.
150   These two new terms are needed for well-authenticated hypophysial connexions.

Ramus anterior
Ramus ascendens
Ramus posterior

LOBUS FRONTALIS
Polus frontalis
Sulcus precentralis
Gyrus precentralis
Gyrus frontalis superior
Sulcus frontalis superior
Gyrus frontalis medius
Sulcus frontalis inferior
Gyrus frontalis inferior
   Pars opercularis (Operculum frontale)
   Pars orbitalis
   Pars triangularis

LOBUS PARIETALIS
Sulcus postcentralis
Gyrus postcentralis
Lobulus parietalis superior
Sulcus intraparietalis
Lobulus parietalis inferior
   Operculum frontoparietale
Gyrus supramarginalis
Gyrus angularis

LOBUS OCCIPITALIS
Polus occipitalis
Sulcus occipitalis transversus
Sulcus lunatus
Incisura preoccipitalis

LOBUS TEMPORALIS
Polus temporalis
Sulci temporales transversi
Gyri temporales transversi
Gyrus temporalis superior
  Operculum temporale
Sulcus temporalis superior
Gyrus temporalis medius
Sulcus temporalis inferior
Gyrus temporalis inferior

LOBUS INSULARIS [INSULA]
Gyri insulae
  Gyri breves insulae
  Gyrus longus insulae
Limen insulae

Sulcus centralis insulae
Sulcus circularis insulae

FACIES MEDIALIS ET INFERIOR HEMISPHERII
Sulcus corporis callosi
Gyrus cinguli [cingulatus]
   Isthmus gyri cinguli [cingulatus]
Sulcus cinguli [cingulatus]
Sulcus subparietalis
Gyrus frontalis medialis
Lobulus paracentralis
Precuneus
Sulcus parietoccipitalis
Cuneus
Sulcus calcarinus
Gyrus dentatus
Sulcus hippocampi [hippocampalis]
Gyrus parahippocampalis
   [G. hippocampi]
    Uncus
Gyrus lingualis
Sulcus collateralis
Sulcus rhinalis
Gyrus occipitotemporalis medialis[151]
Sulcus occipitotemporalis
Gyrus occipitotemporalis lateralis
Gyrus rectus
Sulcus olfactorius
Gyri orbitales
Sulci orbitales
Bulbus olfactorius
   Tractus olfactorius
   Trigonum olfactorium
   Striae olfactoriae medialis et lateralis
Gyri olfactorii medialis et lateralis

Rhinencephalon
Substantia perforata rostalis [anterior]
Stria diagonalis (Broca)
Area subcallosa
Gyrus paraterminalis

CORPUS CALLOSUM
Splenium (corporis callosi)
Truncus corporis callosi
Genu corporis callosi
Rostrum corporis callosi
Radiatio corporis callosi
   Forceps frontalis [minor]

[151] The *Gyrus occipitotemporalis medialis* is continuous with the *Gyrus lingualis* and the *Gyrus parahip-pocampalis.*

A 71

Forceps occipitalis [major]
Tapetum
Indusium griseum
   Stria longitudinalis medialis
   Stria longitudinalis lateralis
Gyrus fasciolaris

Lamina terminalis

Commissura rostralis [anterior]

Fornix
Crus fornicis
Corpus fornicis
Taenia [tenia] fornicis
Columna fornicis
Commissura fornicis

Septum pellucidum
Lamina septi pellucidi
Cavum septi pellucidi
Septum precommissurale

Ventriculus lateralis
Pars centralis
Foramen interventriculare
Cornu frontale [anterius]
Cornu occipitale [posterius]
Cornu temporale [inferius]
Stria terminalis
Lamina affixa
Fissura choroidea
Taenia [tenia] choroidea
Plexus choroideus ventriculi lateralis
Bulbus cornus occipitalis [posterioris]
Calcar avis
Eminentia collateralis
Trigonum collaterale
Hippocampus
   Pes hippocampi
   Alveus hippocampi
   Fimbria hippocampi

Sectiones telencephali
Archaeocortex [Archeo-]
Palaeocortex [Paleo]
Neocortex
Mesocortex

Cortex cerebri
   Lamina molecularis [plexiformis]
   Lamina granularis externa

   Lamina pyramidalis externa
   Lamina granularis interna
   Lamina pyramidalis interna
     [ganglionaris]
   Lamina multiformis
   Neurofibrae tangentiales
     Stria laminae molecularis [plexiformis]
     Stria laminae granularis externa
     Stria laminae granularis interna
     Stria laminae pyramidalis interna
     [ganglionaris]
Fibrae arcuatae cerebri
Cingulum
Fasciculus longitudinalis superior
Fasciculus longitudinalis inferior
Fasciculus uncinatus
Radiatio corporis callosi

Nuclei basales
   Corpus striatum
     Nucleus caudatus
       Caput (nuclei caudati)
       Corpus (nuclei caudati)
       Cauda (nuclei caudati)
     Nucleus lentiformis [lenticularis]
       Putamen
       Lamina medullaris lateralis
       Globus pallidus lateralis
       Lamina medullaris medialis
       Globus pallidus medialis
   Claustrum
   Corpus amygdaloideum
     Area amygdaloidea anterior
     Pars basolateralis
     Pars corticomedialis (olfactoria)
   Capsula extrema
   Capsula externa

Capsula interna
   Crus anterius capsulae internae
     Radiationes thalamicae anteriores
     Tractus frontopontinus
   Genu capsulae internae
     Tractus corticonuclearis
   Crus posterius capsulae internae
     Pars thalamolentiformis
       Fibrae corticospinales
       Fibrae corticorubrales
       Fibrae corticoreticulares
       Fibrae corticothalamicae
       Fibrae thalamoparietales

Radiationes thalamicae centrales
Pars sublentiformis
  Radiatio optica
  Radiatio acustica
  Fibrae corticotectales
  Fibrae temporopontinae
Pars retrolentiformis
  Radiationes thalamicae posteriores
  Fasciculus parieto-occipitopontinus
Corona radiata
Commissura rostralis [anterior]
  Pars anterior
  Pars posterior
Neurofibrae associationes
Neurofibrae commissurales
Neurofibrae projectiones

PARS PERIPHERICA
[Systema Nervosum Periphericum]

Nervus, Nervi
  Endoneurium
  Perineurium
  Epineurium
  Neurofibrae afferentes
  Neurofibrae efferentes
  Neurofibrae somaticae
  Neurofibrae viscerales
Ganglion
  Capsula ganglii
  Stroma ganglii
Ganglia craniospinalia [encephalospinalia]
  (sensorialia)[152]
Ganglia spinalia (sensorialia)[152]
Ganglia sensorialia nn. cranialium
  [encephalici]
Ganglia autonomica [visceralia]
  Neurofibrae preganglionares
  Neurofibrae postganglionares
  Ganglion sympathicum [sympatheticum]
    Ramus communicans albus
    Ramus communicans griseus
  Ganglion parasympathicum
    [parasympatheticum]
Nervi spinales
  Plexus nervorum spinalium

Nervi craniales [encephalici]
  Nuclei nervorum cranialium
    [encephalicorum]
    Nuclei originis
    Nuclei terminationis
Nervus mixtus
  Nervus et rami cutanei
  Nervus et rami articulares
  Nervus et rami musculares
Nervus motorius
Nervus sensorius[152]
Ramus communicans
Nervus et ramus autonomici [viscerales]
  Plexus viscerales et vasculares
    Plexus periarteriales
    Nervi vasorum
Vasa nervorum

NERVI CRANIALES [ENCEPHALICI]

NERVI OLFACTORII (I)[153]

NERVUS OPTICUS (II)

NERVUS OCULOMOTORIUS (III)
  Ramus superior
  Ramus inferior
Ganglion ciliare
  Radix oculomotoria [parasympathica/
    parasympathetica]
  Radix [Ramus] sympathica/sympathetica
  Radix [Ramus] nasociliaris
Nervi ciliares breves

NERVUS TROCHLEARIS (IV)
  Decussatio nervorum trochlearium
    (trochlearis)

NERVUS TRIGEMINUS (V)
  Radix sensoria
    Ganglion trigeminale
  Radix motoria

Nervus ophthalmicus
  Ramus tentorii [meningeus]
  Nervus lacrimalis

---

[152] In all terms employing *sensorialis* the alternative *sensorius* was regarded by some as preferable, being closer to the usual English equivalent—*sensory*.

[153] These Roman numerals provide familiar abbreviations for those who use, for example, the *Nervus cranialis tertius* (III) as a synonym for the preferable *Nervus oculomotorius*. The use of such numerals is to be deprecated.

A 73

Ramus communicans (cum nervo
zygomatico)
Nervus frontalis
Nervus supraorbitalis
Ramus lateralis
Ramus medialis
Nervus supratrochlearis
Nervus nasociliaris
Ramus communicans (cum ganglio
ciliari)
Nervi ciliares longi
Nervus ethmoidalis posterior
Nervus ethmoidalis anterior
Rami nasales (nervus ethmoidalis
anterior)
Rami nasales interni
Rami nasales laterales
Rami nasales mediales
Ramus nasalis externus
Nervus infratrochlearis
Rami palpebrales

Nervus maxillaris
Ramus meningeus (medius)
Rami ganglionares
Ganglion pterygopalatinum (*see* page
A 75, *Nervus intermedius*)
Rami orbitales
Rami nasales posteriores superiores
laterales
Rami nasales posteriores superiores
mediales
Nervus nasopalatinus
Ramus pharyngeus
Nervus palatinus major
Rami nasales posteriores inferiores
Nervi palatini minores
Nervus zygomaticus
Ramus zygomaticotemporalis
Ramus zygomaticofacialis
Nervus infraorbitalis
Nervi alveolares superiores
Rami alveolares superiores posteriores
Ramus alveolaris superior medius
Rami alveolares superiores anteriores
Plexus dentalis superior
Rami dentales superiores
Rami gingivales superiores
Rami palpebrales inferiores

Rami nasales externi
Rami nasales interni
Rami labiales superiores

Nervus mandibularis
Ramus meningeus
Nervus massetericus
Nervi temporales profundi
Nervus pterygoideus lateralis
Nervus pterygoideus medialis
Ganglion oticum (*see* page A 75, *Nervus
glossopharyngeus*)
Ramus communicans (cum nervo
pterygoideo mediali)
Nervus musculi tensoris veli palatini
Nervus musculi tensoris tympani
Nervus buccalis
Nervus auriculotemporalis
Nervus meatus acustici externi
Rami membranae tympani
Rami parotidei
Rami communicantes (cum nervo faciali)
Nervi auriculares anteriores
Rami temporales superficiales
Nervus lingualis
Rami isthmi faucium (Rami fauciales)
Rami communicantes (cum nervo
hypoglosso)
Ramus communicans (cum chorda
tympani)
Nervus sublingualis
Rami linguales
Rami ganglionares
Ganglion submandibulare (*see* below,
Nervus facialis)
Nervus alveolaris inferior
Nervus mylohyoideus
Plexus dentalis inferior
Rami dentales inferiores
Rami gingivales inferiores
Nervus mentalis
Rami mentales
Rami labiales inferiores

NERVUS ABDUCENS (VI)

NERVUS FACIALIS [NERVUS
INTERMEDIOFACIALIS] (VII)[154]

---

[154] These two nerves, through usually separate trunks, form a common trunk. Consequently, some
authorities favour the term *Nervus intermediofacialis* especially since they are, in fact, two radices of
the same cranial nerve.

Geniculum (nervi facialis)
Nervus stapedius
Nervus auricularis posterior
  Ramus occipitalis
  Ramus auricularis
  Ramus digastricus
  Ramus stylohyoideus
  Ramus communicans (cum nervo
    glossopharyngeo)
Plexus intraparotideus
Rami temporales
Rami zygomatici
Rami buccales
Ramus lingualis
Ramus marginalis mandibulae
Ramus colli

NERVUS INTERMEDIUS
Ganglion geniculi [geniculatum]
Ganglion pterygopalatinum (*see* page A 74,
  Nervus maxillaris)
  Nervus canalis pterygoidei (Radix
    facialis)
    Nervus petrosus major
    Nervus petrosus profundus
Chorda tympani
Ramus communicans (cum plexu
    tympanico)
Ramus communicans (cum nervo vago)
Ganglion submandibulare
  Ramus sympathicus [-eticus] (ad
    ganglion submandibulare)
  Rami glandulares
(Glanglion sublinguale)

NERVUS VESTIBULOCOCHLEARIS (VIII)
Radix vestibularis
Radix cochlearis

Nervus vestibularis
  Ganglion vestibulare
    Ramus communicans cochlearis
    Pars rostralis [superior]
    Nervus utriculoampullaris
    Nervus utricularis
    Nervus ampullaris anterior
    Nervus ampullaris lateralis
    Pars caudalis [inferior]
    Nervus ampullaris posterior
    Nervus saccularis (Pars rostralis)

Nervus cochlearis
  Ganglion cochleare [spirale cochleae]

NERVUS GLOSSOPHARYNGEUS (IX)
Ganglion rostralis [superius]
Ganglion caudalis [inferius]
  Nervus tympanicus
  Intumescentia [Ganglion] tympanica[155]
  Plexus tympanicus
    Ramus tubarius [tubalis]
    Nervi caroticotympanici
    Ramus communicans (cum ramo
      auriculari nervi vagi)
    Rami pharyngei [pharyngeales]
    Ramus musculi stylopharyngei
    Ramus sinus carotici
    Rami tonsillares
    Rami linguales
Ganglion oticum (*see* page A 74, *Nervus*
    mandibularis)
  Nervus petrosus minor
  Ramus communicans (cum ramo
    meningeo)
  Ramus communicans (cum nervo
    auriculotemporali)
  Ramus communicans (cum chorda
    tympani)

NERVUS VAGUS (X)
Ganglion rostralis [superius]
Ganglion caudalis [inferius]
Ramus meningeus
Ramus auricularis
  Ramus communicans (cum nervo
    glossopharyngeo)
Rami pharyngei [pharyngeales]
Plexus pharyngeus
Rami cardiaci cervicales superiores
Nervus laryngeus superior
  Ramus externus
  Ramus internus
    Ramus communicans (cum nervo
      laryngeo inferiori)
Rami cardiaci cervicales inferiores
Nervus laryngeus recurrens
  Rami tracheales
  Ramio esophagei [eso—]
  Nervus laryngeus inferior
    Ramus communicans (cum ramo
      laryngeo interno)

[155] Formerly *Ganglion tympanicum*, but is not a ganglion in the neural sense.

NOMINA ANATOMICA

Rami cardiaci thoracici
Rami bronchiales
Plexus pulmonalis
Plexus oesophageus [eso—]
Truncus vagalis anterior
Truncus vagalis posterior
Rami gastrici anteriores
Rami gastrici posteriores
Rami hepatici
Rami coeliaci [celiaci]
Rami renales

NERVUS ACCESSORIUS (XI)
Radices craniales [Pars vagalis]
Radices spinales [Pars spinalis]
Truncus nervi accessorii[156]
  Ramus internus
  Ramus externus
    Rami musculares

NERVUS HYPOGLOSSUS (XII)
Rami linguales

NERVI SPINALES

Fila radicularia
Radix ventralis [anterior] [motoria]
Radix dorsalis [posterior] [sensoria]
  Ganglion spinale [sensorius]
Truncus nervi spinalis
Ramus ventralis [anterior]
Ramus dorsalis [posterior]
Rami communicantes
Ramus meningeus
Cauda equina

NERVI CERVICALES

Rami dorsales
  Ramus medialis
  Ramus lateralis
Nervus suboccipitalis
Nervus occipitalis major
Nervus occipitalis tertius

Rami ventrales

PLEXUS CERVICALS
Ansa cervicalis[157]
  Radix superior[158]
  Radix inferior[158]
  Ramus thyrohyoideus
Nervus occipitalis minor
Nervus auricularis magnus
  Ramus posterior
  Ramus anterior
Nervus transversus colli
  Rami superiores
  Rami inferiores
Nervi supraclaviculares
  Nervi supraclaviculares mediales
  Nervi supraclaviculares intermedii
  Nervi supraclaviculares laterales
    [posteriores]

Nervus phrenicus
Ramus pericardiacus
Rami phrenicoabdominales
(Nervi phrenici accessorii)

PLEXUS BRACHIALIS
Trunci plexus
Truncus superior
Truncus medius
Truncus inferior
  Divisiones ventrales [anteriores]
  Divisiones dorsales [posteriores]

Pars supraclavicularis
Nervus dorsalis scapulae
Nervus thoracicus longus
Nervus subclavius
Nervus suprascapularis

Pars infraclavicularis
Fasciculus lateralis
Fasciculus medialis
Fasciculus posterior
Nervus pectoralis medialis
Nervus pectoralis lateralis

---

[156] The *rami internus et externus* are the terminal branches of the *truncus* formed by the *radices*, the *ramus internus* joining the vagal nerve, the *ramus externus* supplying the trapezius and sternocleido-mastoid muscles.

[157] *Ansa cervicalis*. This was formerly the "*Ansa hypoglossi*" but it is derived from the first, second and third cervical nerves.

[158] *Radix superior*. In this and the following term *radix* was preferred to *ramus* because they are the roots and not branches of the *ansa cervicalis*.

Nervus musculocutaneus
　Rami musculares
　Nervus cutaneous antebrachii lateralis
Nervus cutaneous brachii medialis
Nervus cutaneous antebrachii medialis
　Ramus anterior
　Ramus posterior
Nervus medianus
　Radix medialis
　Radix lateralis
　Nervus interosseus [antebrachii] anterior
　Rami musculares
　Ramus palmaris nervi mediani
　Ramus communicans cum nervo ulnari
　Nervi digitales palmares communes
　　Nervi digitales palmares proprii
Nervus ulnaris
　Rami musculares
　Ramus dorsalis nervi ulnaris
　　Nervi digitales dorsales
　Ramus palmaris nervi ulnaris
　Ramus superficialis
　　Nervi digitales palmares communes
　　　Nervi digitales palmares proprii
　Ramus profundus
Nervus radialis
　Nervus cutaneus brachii posterior
　Nervus cutaneus brachii lateralis inferior
　Nervus cutaneous antebrachii posterior
　Rami musculares
　Ramus profundus
　　Nervus interosseus [antebrachii]
　　posterior
　Ramus superficialis
　　Ramus communicans ulnaris
　　Nervi digitales dorsales
Nervi subscapulares
Nervus thoracodorsalis
Nervus axillaris
　Rami musculares
　Nervus cutaneous brachii lateralis
　superior

NERVI THORACICI

Rami dorsales
　Ramus cutaneus lateralis
　Ramus cutaneus medialis

Rami ventrales [Nervi intercostales]
　Ramus cutaneus lateralis [pectoralis/
　　abdominalis]

　Rami mammarii laterales
Nervus intercostobrachiales
　Ramus cutaneous anterior [pectoralis/
　　abdominalis]
　　Rami mammarii mediales
Nervus subcostalis

NERVI LUMBALES [LUMBARES]

Rami dorsales
　Ramus medialis
　Ramus lateralis
　　Nervi clunium superiores

Rami ventrales

NERVI SACRALES ET NERVUS COCCYGEUS

Rami dorsales
　Ramus medialis
　Ramus lateralis
　　Nervi clunium medii

Rami ventrales

PLEXUS LUMBOSACRALIS
Truncus lumbosacralis

Plexus lumbalis [lumbaris]
Nervus iliohypogastricus
　Ramus cutaneus lateralis
　Ramus cutaneus anterior
Nervus ilio-inguinalis
　Nervi scrotales anteriores
　Nervi labiales anteriores
Nervus genitofemoralis
　Ramus genitalis
　Ramus femoralis
Nervus cutaneus femoris lateralis
Nervus obturatorius
　Ramus anterior
　　Ramus cutaneus
　Ramus posterior
　Rami musculares
Nervus obturatorius accessorius
Nervus femoralis
　Rami musculares
　Rami cutanei anteriores
Nervus saphenus
　Ramus infrapatellaris
　Rami cutanei cruris mediales

Plexus sacralis
Nervus obturatorius internus
Nervus piriformis
Nervus m. quadrati femoris
Nervus gluteus superior
Nervus gluteus inferior
Nervus cutaneus femoris posterior
 Nervi clunium inferiores
 Rami perineales
Nervus ischiadicus [sciaticus]
Nervus fibularis [peroneus]
 Nervus fibularis [peroneus] communis
  Nervus cutaneus surae lateralis
  Ramus communicans fibularis
   [peroneus]
Nervus fibularis [peroneus] superficialis
 Rami musculares
 Nervus cutaneus dorsalis medialis
 Nervus cutaneus dorsalis intermedius
  Nervi digitales dorsales pedis
Nervus fibularis [peroneus] profundus
 Rami musculares
 Nervi digitales dorsales, (hallucis lateralis
  et digiti secundi medialis)
Nervus tibialis
 Rami musculares
 Nervus interosseus cruris
 Nervous cutaneus surae medialis
 Nervus suralis
  Nervus cutaneus dorsalis lateralis[159]
  Rami calcanei laterales
 Rami calcanei mediales
 Nervus plantaris medialis
 Nervi digitales plantares communes
  Nervi digitales plantares proprii
 Nervus plantaris lateralis
  Ramus superficialis
   Nervi digitales plantares communes
   Nervi digitales plantares proprii
  Ramus profundus
Nervus pudendus
 Nervi rectales [anapes] inferiores
 Nervi perineales[160]
  Nervi scrotales/labiales posteriores
 Rami musculares
Nervus dorsalis penis

Nervus dorsalis clitoridis

Nervus coccygeus
 Plexus coccygeus
 Nervi anococcygei

PARS AUTONOMICA
[Systema Nervosum Autonomicum][161]

Plexus autonomici [viscerales][162]
Ganglia plexuum autonomicorum
 [visceralium]
Pars thoracica systematis autonomica

PLEXUS AORTICUS THORACICUS
Plexus cardiacus
Ganglia cardiaca
Plexus oesophagealis [eso-]
Rami pulmonales
 Plexus pulmonalis
Pars abdominalis systematis autonomicae

PLEXUS AORTICUS ABDOMINALIS
Plexus coeliacus (celiacus)
Ganglia coeliaca (celiaca)
Ganglia aorticorenalia
Ganglion mesentericum superius
Plexus intermesentericus
Ganglion mesentericum inferius
Ganglia phrenica
Plexus hepaticus
Plexus splenicus [lienalis]
Plexus gastrici
Plexus pancreaticus
Plexus suprarenalis
Plexus renalis
Ganglia renalia
Plexus uretericus
Plexus testicularis
Plexus ovaricus
Plexus mesentericus superior
Plexus mesentericus inferior
Plexus rectalis superior
Plexus entericus
 Plexus subserosus

---

[159] This is really the continuation of the *N. suralis.*
[160] These nerves usually have identifiable medial and lateral branches, which may be so named if desired.
[161] For the nuclei and fibre tracts of the S.N.A. *see* "Pars centralis" of the nervous system.
[162] The synonym *visceralis* for *autonomicus* has been added in response to numerous requests. Only the larger perivascular autonomic plexuses have been named. Since they are invariably named after vessels which they accompany, the manufacture of terms for smaller plexuses is a simple process.

Plexus myentericus
Plexus submucosus
Plexus iliaci
Plexus femoralis
Pars pelvica systematis autonomica

PLEXUS HYPOGASTRICUS SUPERIOR [NERVUS
    PRESACRALIS][163]
  Nervus hypogastricus [dexter/sinister]
Plexus hypogastricus inferior [Plexus
    pelvicus/pelvinus][163]
Plexus rectales medii
Plexus rectales inferiores
Plexus prostaticus
Plexus deferentialis
Plexus uterovaginalis
  Nervi vaginales
Plexus vesicales
Nervi cavernosi penis
Nervi cavernosi clitoridis

PARS SYMPATHICA [—ETICA]

TRUNCUS SYMPATHICUS [—ETICUS]
Ganglia trunci sympathici
Rami interganglionares
Rami communicantes
Ganglia intermedia

Ganglion cervicale superius
  Nervus jugularis
  Nervus caroticus internus
  Plexus caroticus internus
  Nervi carotici externi
  Plexus caroticus externus
  Plexus caroticus communis
  Rami laryngopharyngei
  Nervus cardiacus cervicalis superior

Ganglion cervicale medium
  Ganglion vertebrale
  Nervus cardiacus cervicalis medius

Ganglion cervicothoracicum (stellatum)[164]
  Ansa subclavia
  Nervus cardiacus cervicalis inferior
  Plexus subclavius
  Nervus vertebralis
  Plexus vertebralis

Ganglia thoracica
  Nervi cardiaci thoracici
  Nervus splanchnicus thoracicus major
  Ganglion thoracicus splanchnicum
  Nervus splanchnicus thoracicus minor
  Ramus renalis
  (Nervus splanchnicus thoracicus imus)

Ganglia lumbalia [lumbaria]
  Nervi splanchnici lumbales [lumbares]

Ganglia sacralia
  Nervi splanchnici sacrales
  Ganglion impar

PARS PARASYMPATHICA [—ETICA]
Nervus terminalis
Ganglion terminale
Ganglion ciliare (*see* page A 73)
Ganglion pterygopalatinum (*see* page A 75)
Ganglion oticum (*see* pages A 74, 75)
Ganglion submandibulare (*see* page A 75)
Ganglion sublinguale (*see* page A 75)
Nervi splanchnici pelvici [Nervi erigentes]
Ganglia pelvica

ORGANA SENSUUM [SENSORIA][165]

ORGANUM VISUS [VISUALE]

OCULUS

Nervus opticus

---

[163] The superior part of the hypogastric plexus is, of course, a median, singular structure, whereas the inferior part is bilateral. Some authorities regard the superior part as dividing immediately into right and left inferior hypogastric plexuses. Here, an interconnecting strand (itself plexiform) is regarded as intervening between the superior and inferior hypogastric plexuses, and the strand is named *N. hypogastricus*.

[164] The inferior cervical ganglion is fused with the first (and occasionally also the second) thoracic ganglion to form the familiar *Ganglion stellatum* of clinical medicine in 75–80% of people.

[165] Many histological terms which appeared in this section in the third edition have been transferred to *Nomina Histologica*.

Pars intracranialis
Pars intracanalicularis
Pars orbitalis
Pars intraocularis
  Pars postlaminaris
  Pars intralaminaris
  Pars prelaminaris
Vagina externa nervi optici
Vagina interna nervi optici
  Spatia intervaginalia

BULBUS OCULI

Polus anterior
Polus posterior
Equator
Meridiani
Axis bulbi externus
Axis bulbi internus
Axis opticus

TUNICA FIBROSA BULBI

Sclera
  Sulcus sclerae
    Reticulum trabeculare [Lig.
      pectinatum]
      Pars corneoscleralis
      Pars uvealis
  Sinus venosus sclerae
  Lamina episcleralis
  Substantia propria sclerae
  Lamina fusca sclerae
  Lamina cribrosa sclerae

Cornea
  Annulus [Anulus] conjunctivae
  Limbus corneae
  Vertex corneae
  Facies anterior
  Facies posterior
  Epithelium anterius

Lamina limitans anterior[166]
Substantia propria (corneae)
Lamina limitans posterior[166]
Epithelium posterius

TUNICA VASCULOSA BULBI

Choroidea
  Lamina suprachoroidea
  Spatium perichoroideale
  Lamina vasculosa
  Lamina choroidocapillaris
  Complexus [Lamina] basalis

Corpus ciliare
  Corona ciliaris
  Processus ciliares
  Plicae ciliares
  Orbiculus ciliaris
  Musculus ciliaris[167]
    Fibrae meridionales
    (Fibrae longitudinales)
    Fibrae radiales
    Fibrae circulares
  Lamina basalis

Iris
  Margo pupillaris
  Margo ciliaris
  Facies anterior
  Facies posterior
  Annulus [Anulus] iridis major
  Annulus [Anulus] iridis minor
  Plicae iridis
  Pupilla
    Musculus sphincter pupillae
    Musculus dilator pupillae
  Stroma iridis
  Epithelium pigmentosum
  Spatia anguli iridocornealis
  Circulus arteriosus iridis major
  Circulus arteriosus iridis minor
  (*Membrana pupillaris*)

---

[166] These are, of course, the anterior membrane of Bowman and the posterior membrane of Descemets. Some histologists wish to retain these eponyms, but the Committee declined to depart from its general policy of avoiding eponymous terms.

[167] Controversy persists regarding classification (by their directions) of the fibres of this muscle. *Fibrae radiales* and *Fibrae longitudinales* have been added, but the latter are, according to some experts, identical with *Fibrae meridionales*. Some authorities recognize four major groups of *fibres* and would change *Musculies ciliaris* to *Musculies quadriceps oculi*, but the Committee rejected this. It is not the responsibility of the I.A.N.C. to dogmatize but only to provide acceptable terminology. A more satisfactory terminology for this muscle must await agreement amongst the experts.

Tunica interna [sensoria] bulbi

Retina
  Pars optica retinae
    Pars pigmentosa
    Pars nervosa
  Ora serrata
  Pars ciliaris retinae
  Pars iridica retinae
  Discus nervi optici
    Excavatio disci
  Macula
    Fovea centralis
    Foveola

Vasa sanguinea retinae
  Circulus vasculosus nervi optici
  Arteriola/Venula temporalis retinae
    superior
  Arteriola/Venula temporalis retinae
    inferior
  Arteriola/Venula nasalis retinae superior
  Arteriola/Venula nasalis retinae inferior
  Arteriola/Venula macularis superior
  Arteriola/Venula macularis inferior
  Arteriola/Venula medialis retinae

Camera anterior bulbi
  Angulus iridocornealis
  Humor aquosus

Camera posterior bulbi
  Humor aquosus

Camera vitrea bulbi
  Corpus vitreum
  (*Arteria hyaloidea*)
  Canalis hyaloideus
  Fossa hyaloidea
  Membrana vitrea
  Stroma vitreum
  Humor vitreus

Lens
  Substantia lentis
  Cortex lentis
  Nucleus lentis
  Fibrae lentis[168]
  Epithelium lentis
  Capsula lentis

Polus anterior lentis
Polus posterior lentis
Facies anterior lentis
Facies posterior lentis
Axis lentis
Equator lentis
Radii lentis
Zonula ciliaris
  Fibrae zonulares
  Spatia zonularia

ORGANA OCULI ACCESSORIA

Musculi bulbi
  Musculus orbitalis
  Musculus rectus superior
  Musculus rectus inferior
  Musculus rectus medialis
  Musculus rectus lateralis
    Lacertus musculi recti lateralis
  Annulus [Anulus] tendineus communis
  Musculus obliquus superior
    Trochlea
    Vagina tendinis musculi obliqui
      superioris
  Musculus obliquus inferior
  Musculus levator palpebrae superioris
    Lamina superficialis
    Lamina profunda

Fasciae orbitales
  Periorbita
  Septum orbitale
  Fasciae musculares
  Vagina bulbi[169]
  Spatium episclerale
  Corpus adiposum orbitae

Supercilium

Palpebrae
  Palpebra superior
  Palpebra inferior
  Facies anterior palpebrarum
  (Plica palpebronasalis)
  Facies posterior palpebrarum
  Rima palpebrarum
  Commissura palpebrarum lateralis
  Commissura palpebrarum medialis

---

[168] A minority of the Committee wished to change *Fibrae lentis* to *Cellulae lentis*.
[169] Tenon's capsule.

Angulus oculi lateralis
Angulus oculi medialis
Limbi palpebrales anteriores
Limbi palpebrales posteriores
Cilia
Tarsus superior
Tarsus inferior
Ligamentum palpebrale mediale
Raphe palpebralis lateralis
Ligamentum palpebrale laterale
Gll. tarsales
Gll. ciliares
Gll. sebaceae
Musculus tarsalis superior
Musculus tarsalis inferior

TUNICA CONJUNCTIVA
Plica semilunaris conjunctivae
Caruncula lacrimalis
Tunica conjunctiva bulbi
Tunica conjunctiva palpebrarum
Fornix conjunctivae superior
Fornix conjunctivae inferior
Saccus conjunctivalis
Glandulae conjunctivales

APPARATUS LACRIMALIS
Glandula lacrimalis
Pars orbitalis
Pars palpebralis
Ductuli excretorii
(Glandulae lacrimales accessoriae)
Rivus lacrimalis
Lacus lacrimalis
Papilla lacrimalis
Punctum lacrimale
Canaliculus lacrimalis
Ampulla canaliculi lacrimalis
Saccus lacrimalis
Fornix sacci lacrimalis
Ductus nasolacrimalis
Plica lacrimalis

ORGANUM VESTIBULOCOCHLEARE

AURIS INTERNA

LABYRINTHUS MEMBRANACEUS
Endolympha
Perilympha

LABYRINTHUS VESTIBULARIS
Ductus endolymphaticus
Saccus endolymphaticus
Ductus utriculosaccularis
Utriculus
Ductus semicirculares
Ductus semicircularis anterior
Ductus semicircularis posterior
Ductus semicircularis lateralis
Membrana propria ductus semicircularis
Membrana basalis ductus semicircularis
Ampullae membranaceae
Ampulla membranacea anterior
Ampulla membranacea posterior
Ampulla membranacea lateralis
Sulcus ampullaris
Crista ampullaris
Cupula
Crura membranacea
Crus membranaceum simplex
Crura membranacea ampullaria
Crus membranaceum commune
Ductus reuniens
Sacculus
Maculae
Macula utriculi
Macula sacculi
Statoconia
Membrana statoconiorum
Endolympha
Perilympha

LABYRINTHUS COCHLEARIS
Spatium perilymphaticum
Scala vestibuli
Scala tympani
Aqueductus vestibuli
Aqueductus cochleae
Ductus cochlearis
Caecum [Cecum] cupulare
Caecum [Cecum] vestibulare
Paries tympanicus ductus cochlearis
[Membrana spiralis]
Organum spirale
Lamina basilaris
Crista spiralis [Ligamentum spirale]
Foramina nervosa
Limbus laminae spiralis osseae
Labium limbi vestibulare
Labium limbi tympanicum
Membrana tectoria
Dentes acustici

Sulcus spiralis internus
Sulcus spiralis externus
Membrana reticularis
Vas spirale
Paries vestibularis ductus cochlearis
    [Membrana vestibularis]
Paries externus ductus cochlearis
Crista basilaris
Prominentia spiralis
Vas prominens
Stria vascularis
Ganglion spirale cochleae

Vasa auris internae
Arteria labyrinthi
  Rami vestibulares
  Ramus cochlearis
  Glomeruli arteriosi cochleae
Venae labyrinthi
  Vena spiralis modioli
  Venae vestibulares
  Vena aqueductus vestibuli
  Vena aqueductus cochleae

LABYRINTHUS OSSEUS

Vestibulum
Recessus sphericus
Recessus ellipticus
Crista vestibuli
Pyramis vestibuli
Recessus cochlearis
Maculae cribrosae
  Macula cribrosa superior
  Macula cribrosa media
  Macula cribrosa inferior

Canales semicirculares ossei
  Canalis semicircularis anterior[170]
  Canalis semicircularis posterior
  Canalis semicircularis lateralis
Ampullae osseae
Ampulla ossea anterior
Ampulla ossea posterior
Ampulla ossea lateralis

Crura ossea
Crus osseum commune
Crus osseum simplex
Crura ossea ampullaria

Cochlea
Cupula cochleae
Basis cochleae
Canalis spiralis cochleae
Modiolus
  Basis modioli
  Lamina modioli
  Canalis spiralis modioli
  Canales longitudinales modioli
Lamina spiralis ossea
  Hamulus laminae spiralis
Helicotrema
Lamina spiralis secundaria

Meatus acusticus internus
Porus acusticus internus
Fundus meatus acustici interni
  Crista transversa
  Area nervi facialis
  Area cochleae
    Tractus spiralis foraminosus
  Area vestibularis superior
  Area vestibularis inferior
    Foramen singulare

AURIS MEDIA[171]

CAVITAS TYMPANICA [CAVUM TYMPANI]
Paries tegmentalis
  Recessus epitympanicus
  Pars cupularis
Paries jugularis
  Prominentia styloidea
Paries labyrinthicus
  Fenestra vestibuli
  Fossula fenestrae vestibuli
  Promontorium
    Sulcus promontorii
    Subiculum promontorii

---

[170]   This is also often still named *Can. s. superior*.
[171]   In the third edition *Auris media* and *Cavum tympani* were regarded as synonymous. The Committee approved the view that the latter is a part of the former. *Auris media* could be conveniently divided into a *Cavitas tympanica, Adnexa mastoidea*, and *Tuba auditiva*, as suggested by the French Commission.

Sinus tympani
Fenestra cochleae
Fossula fenestrae cochleae
Crista fenestrae cochleae
Processus cochleariformis
Membrana tympani secundaria

(ADNEXA MASTOIDEA)
Paries mastoideus
  Antrum mastoideum
  Aditus ad antrum
    Prominentia canalis semicircularis
      lateralis
    Prominentia canalis facialis
  Eminentia pyramidalis
  Fossa incudis
  Sinus posterior
  Apertura tympanica canaliculi chordae
    tympani
  Cellulae mastoideae
  Cellulae tympanicae
Paries caroticus
Paries membranaceus

Membrana tympani
Pars flaccida
Pars tensa
Plica mallearis anterior
Plica mallearis posterior
Prominentia mallearis
Stria mallearis
Umbo membranae tympani
Annulus [Anulus] fibrocartilagineus

Ossicula auditus [*auditoria*]
Stapes
  Caput stapedis
  Crus anterius
  Crus posterius
  Basis stapedis
Incus
  Corpus incudis
  Crus longum
    Processus lenticularis
  Crus breve
Malleus
  Manubrium mallei
  Caput mallei
  Collum mallei
  Processus lateralis
  Processus anterior

Articulationes ossiculorum auditus
  Articulatio incudomallearis
  Articulatio incudostapedia
  Syndesmosis tympanostapedia
Ligamenta ossiculorum auditus
  Ligamentum mallei anterius
  Ligamentum mallei superius
  Ligamentum mallei laterale
  Ligamentum incudis superius
  Ligamentum incudis posterius
  Membrana stapedis
  Ligamentum annulare [anulare] stapedis
Musculi ossiculorum auditus
  Musculus tensor tympani
  Musculus stapedius

Tunica mucosa cavitatis tympani
Plica mallearis posterior
Plica mallearis anterior
Plica chordae tympani
Recessus membranae tympani anterior
Recessus membranae tympani superior
Recessus membranae tympani posterior
Plica incudis
Plica stapedis

TUBA AUDITIVA [AUDITORIA]

Ostium tympanicum tubae auditivae
Pars osseae tubae auditivae
  Isthmus tubae auditivae
  Cellulae pneumaticae

Pars cartilaginea tubae auditivae
  Cartilago tubae auditivae
    Lamina [cartilaginis] medialis
    Lamina [cartilaginis] lateralis
    Lamina membranacea
Tunica mucosa
  Glandulae tubariae
Ostium pharyngeum tubae auditivae

AURIS EXTERNA

MEATUS ACUSTICUS EXTERNUS
Porus acusticus externus
Incisura tympanica
Meatus acusticus externus cartilagineus
  Cartilago meatus acustici

Incisurae cartilaginis meatus acustici
Lamina tragi

AURICULA
Lobulus auriculae
Cartilago auriculae
Helix
  Crus helicis
  Spina helicis
  Cauda helicis
Anthelix
  Fossa triangularis
  Crura anthelicis
Scapha
Concha auriculae
  Cymba conchae
  Cavitas [Cavum] conchae
Antitragus
Tragus
Incisura anterior (auris)
Incisura intertragica
(Tuberculum auriculae)
(Apex auriculae)
Sulcus auriculae posterior
(Tuberculum supratragicum)
Isthmus cartilaginis auris
Incisura terminalis auris
Fissura antitragohelicina
Sulcus anthelicis transversus
Sulcus cruris helicis
Fossa anthelicis
Eminentia conchae
Eminentia scaphae
Eminentia fossae triangularis

Ligamenta auricularia
  Ligamentum auriculare anterius
  Ligamentum auriculare superius
  Ligamentum auriculare posterius

Musculi auriculares
Musculus helicis major
Musculus helicis minor
Musculus tragicus
Musculus pyramidalis auriculae
Musculus antitragicus
Musculus transversus auriculae

Musculus obliquus auriculae
(Musculus incisurae helicis)

ORGANUM OLFACTUS [OLFACTORIUM]

Regio olfactoria tunicae mucosae nasi
Glandulae olfactoriae

ORGANUM GUSTUS [GUSTATORIUM]

Caliculus gustatorius [Gemma gustatoria]
Porus gustatorius

INTEGUMENTUM COMMUNE[172]

CUTIS
Sulci cutis
Cristae cutis
Retinacula cutis
Toruli tactiles
(Foveola coccygea)
Retinaculum caudale

EPIDERMIS

DERMIS [CORIUM][173]
Papillae
Tela subcutanea
  Panniculus adiposus
Terminationes nervorum (*see Nomina Histologica*)
Pili
Lanugo
Capilli
Supercilia
Cilia
Barba
Tragi
Vibrissae
Hirci
Pubes
Folliculus pili
Musculi arrectores pilorum

---

[172] Many terms now appearing in *Nomina Histologica* have been omitted from this section.
[173] Since *Dermis* is in wider usage and is more informative than *Corium*, the relative positions of these synonyms have been reversed. In *derm*atological terminology *corium* is little used to form other words, whereas the stem "derm-" is familiar to all.

Flumina pilorum
Vortices pilorum
Cruces pilorum

UNGUIS
Matrix unguis
Vallum unguis
Corpus unguis
Radix unguis
Lunula
Margo occultus
Margo lateralis
Margo liber
Perionyx
Eponychium
Hyponychium

MAMMA
Papilla mammaria
Corpus mammae

GLANDULA MAMMARIA
Processus lateralis [axillaris]
Lobi glandulae mammariae
Lobuli glandulae mammariae
Ductus lactiferi
Sinus lactiferi
Areola mammae
Glandulae areolares
(Mammae accessoriae)
Ligamenta suspensoria mammaria
Mamma masculina

NOMINA HISTOLOGICA

Approved by the
Eleventh International Congress of Anatomists
and World Association of Veterinary Anatomists
at Mexico City, August 1980

CONTENTS

CONTENTS

CONTENTS

STYLE USAGE

1. Terms within squared brackets are officially recognized as synonyms or alternatives, e.g. *Cellula columnaris [prismatica], Cilium [Microcilium]*, and *Particula elementaria [Spherula membranosa]*. The first example illustrates an alternative for only a part of the complete term.
2. The virgule, or slash, is used for three purposes: to indicate bilateral structures, e.g. *Trigonum fibrosum dextrum/sinistrum*; secondly, to indicate position, e.g. *Epitheliocytus limitans internus/externus*; thirdly, to indicate comparable or homologous structures e.g. *UNGUI-CULA (Car)/UNGULA (Un)*.
3. Terms are placed in rounded brackets for four purposes: firstly, to indicate that the structure named is inconsistent, e.g. (*Tela submucosa*) subordinate to *Bronchus*; secondly, to indicate certain unofficial but important alternatives, e.g. *Neuronum multipolare (sympathicum)*; thirdly, to indicate additional components of certain terms which are frequently omitted, e.g. *Bulbus (pili)*; fourthly, to indicate that some terms are restricted to a particular species, genus, or order. These abbreviations designate the species in which certain structures occur; where no such designation is made, the structure is common to man and these mammals or its distribution is uncertain. Structures unique to man are designated "Homo."

(Car)	—Carnivora	(su)	—Sus scrofa domestica
(Un)	—Ungulata	(b)	—Bos taurus
(Ru)	—Ruminantia	(ov)	—Ovis aries
(fe)	—Felis catus	(cap)	—Capra hircus
(ca)	—Canis familiaris	(eq)	—Equus caballus

PRINCIPLES OF TERM CONSTRUCTION

The following principles served as guides for this new edition. When proposing new terms or revisions, these principles should be followed.

1. The list is a systematic organization of terms in Latin according to body systems. No attempt is made at translation into other languages, but vernacularization of Latin terms is, of course, commonplace and is particularly easy in Romance languages and English, which contain large numbers of words derived from Latin. While it is impossible, and perhaps undesirable to prevent vernacularization, it is often unhelpful to international understanding, a point of special importance in scientific literature, where the official terms *ought* to be used.

2. Because the list is not intended as a complete outline of microscopic anatomy, repetition of terms is omitted wherever convenient.

3. The list is not intended to serve as a dictionary or textbook for these terms. Only terms of too recent an origin to have been included in standard medical dictionaries receive a partial definition by way of footnote.

4. The terms and organization used in the *Nomina Anatomica, Nomina Anatomica Veterinaria,* and *Nomina Embryologica* are used as a basis for additions and revisions.

5. The terms chosen are etymologically related to morphologic features whenever possible.

6. The use of synonyms is avoided wherever possible. New terms proposed as revisions may appear as an acceptable alternative and in a later edition become the primary term. Terms which are descriptive and instructive have been given preference.

7. Eponyms have been excluded. Reference to eponyms and historic information may appear in footnotes.

8. The list is compiled to facilitate communication between anatomists and for others whose communications make use of anatomical terms.

INTRODUCTION TO SECOND EDITION

By invitation of Professor Roger Warwick, Honorary Secretary of the International Anatomical Nomenclature Committee (I.A.N.C.), the Subcommittee for Histology met at the Ciba House, Ciba Foundation in London during June 25th–29th, 1979, to prepare a Second Edition of *Nomina Histologica* to be included in a Fifth Edition of *Nomina Anatomica*. Subcommittee members attending the meeting were Professor M. T. Rakhawy, Professor I. R. Telford, Professor A. F. Weber, and Professor R. Wegmann. Also meeting with the Subcommittee were Professor R. Warwick, Honorary Secretary, members of other I.A.N.C. Subcommittees, Professor A. Tavares de Sousa, Professor T. Donáth, Professor F. Strauss, and Professor H. Werneck, and these members of the International Committee on Veterinary Anatomical Nomenclature: Professor Breazile, Professor Hullinger, and Professor Venable.

Correspondence from many histologists throughout the world was reviewed and their suggestions collated and summarized. Each term in the First Edition was reexamined. Particular attention was given to the incorporation of terms that agreed with the organization and content in the *Nomina Anatomica, Nomina Embryologica, Nomina Histologica Veterinaria* (an unpublished working document), and a proposed *Nomina Neuroanatomica*. The Subcommittee convened with Professor Arey of the Subcommittee on Embryology at the Fourth International Symposium of the Morphological Sciences at the University of Toledo August 1st–5th, 1979. All members of the Subcommittee were apprised by mail of the changes and additions proposed at the London and Toledo meetings for the new Second Edition. Their critiques and suggestions were requested before the next meeting of the Subcommittee.

A formal working session of the Subcommittee was held in Mexico City, August 17th–23rd, 1980, in connection with the 11th International Congress of Anatomists. Further proposals for revision of the final draft were made and approved by the Subcommittee. The Second Edition was then adopted by the International Congress of Anatomists and the World Congress of Veterinary Anatomists at Mexico City.

To fill the vacancy created by the resignation of Professor T. E. Hunt, Professor I. R. Telford was appointed Convenor of the Subcommittee on Histology; Professor R. L. Hullinger was appointed Secretary.

The Subcommittee acknowledges the contribution of all these former members who have resigned: Professor D. Bulmer, Professor F. W. Fyfe, Professor T. E. Hunt, Professor A. G. Knorre, Professor Y. Kopaev, Professor A. N. Studitsky, Professor R. Wegmann, and Professor E. Yamada.

These new members have been added to the Subcommittee: Professor R. L. Hullinger (U.S.A.), Professor M. J. Koering (U.S.A.), Professor V. Monesi (Italy), Professor P. Nieuwenhuis (The Netherlands), Professor W. J. Paule (U.S.A.), Professor I. R. Telford (U.S.A.), Professor A. Tavares de Sousa (Portugal), and Professor O. Wolkova (U.S.S.R.).

As the Second Edition goes to press, we express gratitude to Professor T. E. Hunt, former Subcommittee Convenor, who was so active and so productive in the preparation of the First Edition and to the many colleagues who proposed the additions and improvements in this edition. We invite the continued use and criticism of the list.

Finally, we regret the loss by death, of Professors T. E. Hunt, Burton L. Baker, and Raymund L. Zwemer. Professor Baker was an active and esteemed member of the Subcommittee for many years. As Convener of the Finance Subcommittee, Professor Zwemer was a primary supporter of the Subcommittee and member of the Main Committee.

Professor IRA R. TELFORD
Convenor

Professor RONALD L. HULLINGER
Secretary

Subcommittee on Histology
International Anatomical Nomenclature Committee

INTERNATIONAL ANATOMICAL NOMENCLATURE COMMITTEE
HISTOLOGY SUBCOMMITTEE
(see pages A2–A8 for complete address.)

PROF. A. BAIRATI
PROF. MARY DYSON
PROF. R. L. HULLINGER (Secretary)*
PROF. M. J. KOERING
PROF. LATA N. MEHTA
PROF. V. MONESI*
PROF. P. NIEUWENHUIS*
Prof. W. J. Paule
PROF. M. T. RAKHAWY
PROF. I. R. TELFORD (Convener)*
PROF. A. F. WEBER
PROF. O. WOLKOVA*

* Appointed since the First Edition.

CYTOLOGIA

Polus cellularis
Axis cellularis
Basis cellularis
Apex cellularis

CELLULA [Cyto-, -cytus][1]

Cellula amoeboidea
Cellula cuboidea
Cellula columnaris [prismatica]
Cellula dendriformis
Cellula fusiformis
Cellula ovoidea
Cellula polyhedralis
Cellula pyramidalis
Cellula spherica
Cellula squamosa [plana]
Cellula stellata
Invaginatio cellularis
Vesicula superficialis [Caveola]
Processus cellularis
  Processus amoeboideus
  Processus digitiformis
  Processus filiformis
  Processus lamellosus
  Processus polypoideus
Microvillus[2]
Cilium [Microcilium][2]
  Pars extracellularis
  Pars intracellularis
  Filamentum axiale
  Microtubulus centralis
  Diplomicrotubulus periphericus
  Corpusculum basale
    Radix basalis
    Pes basalis
    Triplomicrotubulus
Flagellum

PROTOPLASMA

CYTOPLASMA

Membrana cellularis[3]
  Lamina externa
  Lamina intermedia
  Lamina interna
  Superficies plasmica[4]
  Superficies extraplasmica[4]
  Facies plasmica[4]
  Facies extraplasmica[4]
Plasmalemma[3]
  Glycocalyx
  Exoplasma
  Endoplasma

ORGANELLAE ET INCLUSIONES
CYTOPLASMICAE[5]

Cytocentrum
  Centriolum
  Diplosoma
    Triplomicrotubulus
Mitochondrion
  Membrana mitochondrialis
    Membrana mitochondrialis externa
    Spatium intermembranosum
    Membrana mitochondrialis interna[6]

---

[1] In this edition the Committee chose to limit the use of *Cellula* to the more general case and utilize the ending *-cytus* when forming the specific noun, e.g. *Cellula fusiformis* (page H 3) is retained, but *Cellula tendinea* (page H 8) becomes *Tendinocytus*.

[2] Both the *Microvillus* and the *Cilium* exhibit movement, but uniform, metachronic movement is characteristic of cilia and absent in microvilli. The earlier designation of kinetocilium and stereocilium for structures of generally similar shape, but mobile and immobile respectively, now is inappropriate. A "stereocilium" is an elongated, often branched, microvillus.

[3] *Membrana cellularis* designates the general membrane ("unit membane") of the membrane complex within and limiting a cell. *Plasmalemma* designates the outer trilaminar membrane of the cell.

[4] In deference to methods of microscopy employing freeze fracture techniques, these terms are added for natural surfaces and fracture faces of cytomembranes (*see* Branton et al., 1975, *Science* 190:54).

[5] The classification of some structures as organelles or inclusions remains problematic. Information provided by newer techniques changes this classification. Consequently, they are again in this edition grouped under one heading.

[6] The projections of inner mitochondrial membrane exist as one or a combination of forms depending upon cell type and functional activity. The lamelliform, crescentic type is more widespread, but in steroidogenic cells the tubular and vesicular projections are found.

Crista
Tubulus
Vesicula
Particula elementaria [Spherula
 membranae]
Matrix mitochondrialis
Chromosoma mitochondrialis
Filamentum mitochondriale
Granulum mitochondriale
Inclusio mitochondrialis
Complexus golgiensis [Apparatus
 reticulatus internus][7]
Sacculus lamelliformis
Vesicula
Reticulum endoplasmicum
 [cytoplasmaticum]
Reticulum endoplasmicum
 granulosum
Reticulum endoplasmicum
 nongranulosum
 Cisterna
 Lamella
 Sacculus
 Tubulus
Ribosoma
Polyribosoma
Lamella annulata [anulata]
Vesicula cytoplasmica
Vesicula pinocytotica
Peroxysoma
Lysosoma
Phagosoma
Phagolysosoma [Heterophagosoma]
Autophagosoma
Corpus multitubulare[8]
Corpus multivesiculare
Premelanosoma
Melanosoma
 Corpusculum lamellosum melaniferum
Microtubulus
Microfibrilla
Microfilamentum[9]
Granulum cellulare

Granulum glycogeni
Granulum proteini
Granulum pigmenti
 Granulum carotenoidi
 Granulum ferritini
 Granulum hemosiderini
 Granulum hematoidini
 Granulum lipochromi
 Granulum lipofuscini
 Granulum melanini
 Granulum secretorium
Gutta adipis [Adiposoma]
Inclusio crystalloidea
Corpusculum residuale

NUCLEUS

Nucleus annularis [anularis]
Nucleus bacilliformis
Nucleus fusiformis
Nucleus moniliformis
Nucleus ovoideus
Nucleus piriformis
Nucleus planus
Nucleus polymorphus
Nucleus reniformis
Nucleus segmentalis
Nucleus sphericus
Nucleus vesicularis
Nucleolemma[10]
 Membrana nuclearis externa
 Membrana nuclearis interna
 Cisterna nucleolemmae
 Complexus pori
 Porus nuclearis
 Annulus [Anulus] pori
 Diaphragma pori
Nucleoplasma
 Chromatinum
 Euchromatinum
 Heterochromatinum
 Granulum chromatini

---

[7] For the first edition of *Nomina Histologica* the principle rejecting eponyms was modified in three cases by creating these adjectival eponyms: *golgiensis, purkinjiensis* and *panethensis*. The Committee retained the first and dropped the latter two in the present edition. The term *Complexus sacculorum* has been proposed as an alternative for the first term.

[8] This organelle of the *Endotheliocytus* has also been called a "Weibel-Palade body."

[9] A network of microfilaments has been demonstrated in most cells, related to but of different composition than myofilaments, neurofilaments, or keratofilaments. The network is called "cytoskeleton."

[10] *Nucleolemma* replaces *Karyotheca* of the first edition.

Corpusculum chromatini sexualis[11]
Nucleoplasma filamentosum
Nucleoplasma punctatum
Nucleolus
  Nucleolus principalis
    Nucleolonema
      Pars filamentosa
      Pars granulosa
  Nucleolus accessorius
  Nucleolus compositus
Corpusculum nucleare[12]
Nucleus pyknoticus

DIVISIO NUCLEARIS ET CELLULARIS

Cyclus cellularis
  Periodus mitoticus[13]
    Cellula mitotica
    Nucleus mitoticus
  Periodus intermitoticus[13]
    Cellula intermitotica [interphasica]
    Nucleus intermitoticus
      [interphasicus]
    Intervallum ambiguum
    Intervallum postmitoticum
    Synthesis genomica
    Intervallum premitoticum

MITOSIS [CYCLUS MITOTICUS]

Cytokinesis
Nucleokinesis
Prophasis
  Glomus compactum
  Glomus dispersum
Metaphasis
  Lamina equatorialis
Anaphasis
Telophasis
  Constrictio cytoplasmatica
  Corpusculum intermedium
  Relictum fusi

Corpusculum residuale
Chromosoma
  Crus chromosomatis
  Centromerus [Kinetochorus]
    Centromerus comitans
    Spherula centromeri
  Chromonema
    Pars euchromatica
    Pars heterochromatica
      Pars heterochromatica centralis
  Chromomerus
  Matrix chromosomatis
Chromosoma acrocentricum
Chromosoma dicentricum
Chromosoma metacentricum
Chromosoma monocentricum
Chromosoma polycentricum
Chromosoma submetacentricum
Chromosoma telocentricum
Telomerus
Satelles chromosomalis
  Filum satellitis
Chromosoma satellitiferum
Autosoma
Gonosoma
  Gonosoma femininum
    [X-chromosoma]
  Gonosoma masculinum
    [Y-chromosoma]
Chromosoma nucleolare
Heterosoma
Chromosoma maternum
Chromosoma filiale
Apparatus mitoticus
Radiatio polaris
Centriolum filiale
Fusus mitoticus
  Apparatus fusalis
  Microtubulus fusalis
    Microtubulus chromosomaticus
    Microtubulus continuus
  Equator fusi
Mitosis multipolaris
Diaster
Aster
Amitosis[14]

[11] These nuclear elements have also been called "Barr bodies."
[12] These are characteristic of active cells and may represent a segregation of nucleolar components.
[13] *Periodus mitoticux* has been abbreviated as M and the phases of the *Periodus intermitoticus* abbreviated *Intervallum ambiguum* as G_0, *Intervallum postmitoticum* as G_1, *Synthesis genomica* as S, and *Intervallum premitoticum* as G_2.
[14] There remains disagreement regarding the existence of *Amitosis*.

MEIOSIS/CYCLUS MEIOTICUS

Prophasis I
  Proleptonema
  Leptonema
  Zygonema
    Conjugatio
    Decussatio[15]
  Pachynema
  Chromosoma meioticum
  Chromosoma homologum
  Synapsis
    Complexus synaptonematicus
  Chromosoma bivalens
  Chromosoma quadrivalens
  Diplonema
  Diakinesis
Prometaphasis I
Metaphasis I
Anaphasis I
Telophasis I
Interkinesis
Prometaphasis II
Metaphasis II
Anaphasis II
Telophasis II
Endomitosis
Haploidea
Diploidea
Triploidea
Tetraploidea
Aneuploidea
Polyploidea

JUNCTIONES INTERCELLULARES

Junctio intercellularis simplex
  Junctio intercellularis denticulata
  Junctio intercellularis digitiformis
Junctio intercellularis complex

Macula adherens [Desmosoma]
Hemidesmosoma
Zonula occludens
Zonula adherens
Nexus [Macula communicans][16]
Tonofibrilla
Tonofilamentum
Substantia intercellularis
Spatium intercellulare

HISTOLOGIA GENERALIS

TEXTUS EPITHELIALIS

Cellulae epitheliales[17]
  Epitheliocytus squamosus
  Epitheliocytus cuboideus
  Epitheliocytus columnaris
  Epitheliocytus ciliatus
  Epitheliocytus flagellatus
  Epitheliocytus microplicatus
  Epitheliocytus microvillosus
    Limbus penicillatus
    Limbus striatus
  Epitheliocytus pigmentosus
  Epitheliocytus secretorius
    [Glandulocytus]
  Epitheliocytus neurosensorius[18]
  Epitheliocytus sensorius[18]
Membrana basalis
  Lamina lucida
  Lamina densa [basalis]
  Lamina fibroreticularis

EPITHELIUM SUPERFICIALE

Epithelium simplex squamosum
  Endothelium[19]

---

[15] *Decussatio* connotes "crossing over."

[16] *Nexus [Macula communicans]* is also called the "gap junction" which does not translate readily into Latin.

[17] In this edition, the Committee chose to limit the use of *cellulae* to headings for groups of cells. Specific cells are named using the ending *-cytus* to form the noun.

[18] *Epitheliocytus neurosensorius* is a nerve cell modified as a receptor for sight or olfaction. *Epitheliocytus sensorius* is a non-neuronal cell modified for reception of specific stimuli as in the organs for hearing, equilibration, and taste.

[19] Endotheliocytes and mesotheliocytes are the principal component of endothelia and mesothelia respectively. *Endothelium* is a tissue composed of a single layer, generally of flattened cells derived from mesoderm and lining the vascular system. A tissue of similar structure and origin but which covers the serous membranes is termed *Mesothelium*.

Endotheliocytus
  Vesicula superficialis [Caveola]
  Corpus multitubulare[20]
Mesothelium[19]
  Mesotheliocytus
Epithelium simplex cuboideum
Epithelium simplex columnare
Epithelium pseudostratificatum
  Epitheliocytus superficialis
  Epitheliocytus intercalatus
  Epitheliocytus basalis
Epithelium stratificatum squamosum
    cornificatum
Epithelium stratificatum squamosum
    noncornificatum
  Stratum superficiale
  Stratum intermedium [spinosum]
  Stratum basale
Epithelium stratificatum cuboideum
Epithelium stratificatum columnare
Epithelium transitionale[21]
  Stratum superficiale
  Stratum intermedium
  Stratum basale
Epithelium ciliatum
Epithelium sensorium (see ORGANA
    SENSUUM page H 33)

EPITHELIUM GLANDULARE

Glandulocytus
Glandula exocrina
  Exocrinocytus
    Granulum secretorium
      Granulum mucigeni
      Granulum zymogeni
Glandula endocrina
  Endocrinocytus
Glandula intraepithelialis
Glandula unicellularis
  Exocrinocytus caliciformis[22]
Glandula multicellularis

GLANDULA UT ORGANUM[23]

Stroma glandulare
  Capsula glandularis
  Septum interlobare
  Septum interlobulare
  Trabecula glandulae
  Interstitium glandulae
Parenchyma glandulare
  Lobus glandularis
  Lobulus glandularis
Ostium glandulae
Isthmus glandulae
Collum glandulae
Corpus glandulae
Fundus glandulae
Glandula acinosa
Glandula alveolaris
Glandula tubulosa
Glandula tubuloacinosa
Glandula tubuloalveolaris
Glandula simplex
Glandula composita
Glandula ramosa
Portio terminalis
  Acinus
  Alveolus
  Tubulus
  Tubuloacinus
  Tubuloalveolus
  Semiluna serosa
Canaliculus intracellularis
Canaliculus intercellularis
Ductus secretorius
Ductus intralobularis
  Ductus striatus
  Ductus intercalatus
Ductus interlobularis
Ductus interlobaris
Ductus excretorius [glandulae]
Myoepitheliocytus fusiformis
Myoepitheliocytus stellatus
Glandula serosa

---

[20] This organelle of the *Endotheliocytus* has also been called a "Weibel-Palade body."
[21] This epithelium, characteristic of but not limited to the urinary bladder and urethra, is described by some investigators (Petry and Amon, 1966 Z. Zellforsch, 69:587) as pseudostratified cuboidal or columnar epithelium depending on the degree of distension of the organ.
[22] *Exocrinocytus caliciformis* is commonly referred to as a "goblet cell."
[23] In the first edition of *Nomina Histologica* terms now under the new heading of GLANDULA UT ORGANUM were included under EPITHELIUM GLANDULARE.

Glandula mucosa
Glandula seromucosa
Glandula merocrina
Glandula apocrina
Glandula holocrina
Glandula exocrina
Glandula endocrina

TEXTUS CONNECTIVUS

Cellulae textus connectivi
    Fibroblastocytus[24]
    Fibrocytus
    Reticulocytus[25]
    Pericytus [Periangiocytus]
    Lymphocytus
    Macrophagocytus[26]
        Macrophagocytus stabilis
        Macrophagocytus nomadicus
    Granulocytus basophilus textus[27]
    Leucocytus globularis
    Plasmocytus
    Adipocytus uniguttularis
    Adipocytus multiguttularis
Cellula pigmentosa
    Chromatophorocytus
    Melanophorocytus
    Hemosiderophorocytus
    Lipochromophorocytus
Substantia intercellularis
    Fibrae textus connectivi
        Fibra collagenosa
            Fasciculus (collagenosus)
            Fibrilla (collagenosa)
            Protofibrilla (collagenosa)
            Microfibrilla (collagenoidea)

Fibra elastica
    Pars amorpha (elastica)
    Pars filamentosa (elastica)
    Rete elasticum
    Lamella elastica
    Membrana elastica fenestrata
Fibra reticularis
Substantia fundamentalis [amorpha]
Textus connectivus collagenosus[28]
    Textus connectivus collagenosus laxus
    Textus connectivus collagenosus
        compactus
    Textus connectivus collagenosus
        compactus regularis
    Textus connectivus collagenosus
        compactus irregularis
Textus connectivus elasticus
Textus connectivus reticularis
Textus adiposus albus
Textus adiposus fuscus
Textus connectivus pigmentosus

TENDO

Fasciculus tendineus
Fibra tendinea
Tendinocytus
Endotendineum
Peritendineum
Epitendineum
Vagina synovialis tendinis
    Pars parietalis
    Pars tendinea[29]
    Stratum fibrosum
    Stratum synoviale
    Mesotendineum

---

[24] It is common that these terms are shortened, i.e. "Fibroblastus" and "Macrophagus."

[25] The Committee chose to limit the use of *Reticulocytus* to the stem cell of reticular connecting tissue. *Reticulocytus* has been commonly used for both the multipotent, less differentiated, stem cell of reticular connective tissue and the maturing, anucleate erythrocyte. In the first edition of *Nomina Histologica* this ambiguity was addressed by using *Cellula reticularis* and *Reticulocytus* respectively (*see* footnote 34).

[26] Macrophages of various locations have, as a consequence of their similar structure and function, been grouped and called "macrophage system," "mononuclear phagocyte system," or formerly "reticuloendothelial system." The *Macrophagocytus stabilis* is also called "histiocyte," which, by etymology, is rather unsatisfactory.

[27] It is generally accepted that the basophilic granulocyte enters the tissue and becomes the "mastocytus."

[28] Fibrous connective tissues are classified according to the most prevalent fibre component. In nearly all classifications there will also be some of the other fibres, e.g. collagneous connecting tissue also contains fibres of reticulin and elastin.

[29] *Pars tendinae* replaces *Pars visceralis* of the first edition; "visceral" is inappropriately applied to tendon.

Cavitas synovialis
Bursa synovialis
  Stratum fibrosum
  Stratum synoviale
Junctio myotendinea

LIGAMENTUM ET FASCIA[30]

Ligamentum fibrarum collagenosarum
Ligamentum fibrarum elasticarum
Fibra elastica
Fibra collagenosa

TEXTUS CARTILAGINEUS

Chondroblastocytus
Chondrocytus
Aggregatio chondrocytica
Matrix cartilaginea
  Substantia fundamentalis
  Fibra matricis
  Matrix territorialis cellularum
  Matrix interterritorialis
Lacuna cartilaginea
Cartilago hyalina
Cartilago fibrosa [collagenosa]
Cartilago elastica
Perichondrium
  Stratum fibrosum
  Stratum chondrogenicum[31]
Canalis cartilaginis

CHONDROHISTOGENESIS

Textus prechondrialis
  Chondroblastocytus

TEXTUS OSSEUS

Osteoblastocytus
Osteocytus
Osteoclastocytus
Matrix ossea
  Substantia fundamentalis

Fibra collagena
Crystallum hydroxyapatiti
Canaliculus osseus
Lacuna ossea
  Lacuna osteocyti
  Lacuna erosionis
Textus osseus reticulofibrosus
Textus osseus lamellaris

OSTEOHISTOGENESIS

Textus osteogenicus
  Osteoblastocytus
Textus osteoideus
Centrum ossificationis
Crystallum hydroxyapatiti
Trabecula ossea
Osteoclastocytus
Lacuna erosionis

OS UT ORGANUM[32]

Textus osseus
Periosteum
  Stratum fibrosum
  Stratum osteogenicum
  Fibra perforans
Endosteum
Os compactum
  Lamella ossea
    Lamella circumferentialis externa
    Lamella circumferentialis interna
    Lamella interstitialis
  Osteonum
    Lamella osteoni
    Canalis centralis
    Canalis perforans
  Linea cementalis
  Linea resorptionis
Os spongiosum [trabeculare]
  Trabecula ossea
Canalis nutricius
Cavitas medullaris
Medulla ossium flava
Medulla ossium rubra
Epiphysis

[30] These terms added in this edition.
[31] *Stratum chondrogeneticum* replaces "stratum cellulare" of the first edition.
[32] In the first edition of *Nomina Histologica* terms now under the new heading of OS UT ORGANUM were included under **TEXTUS OSSEUS**.

H 9

NOMINA HISTOLOGICA

Diaphysis
Metaphysis

OSTEOGENESIS

Textus osteogenicus
Textus osteoideus
Os membranaceum
Os cartilagineum

OSTEOGENESIS MEMBRANACEA

Centrum ossificationis
Os membranaceum reticulofibrosum
    [primarium]
  Trabecula ossea
Os membranaceum lamellosum
    [secúndarium]
  Lamella ossea
Osteoblastocytus

OSTEOGENESIS CARTILAGINEA

Ossificatio perichondrialis
  Perichondrium
    Stratum osteogenicum
      Osteoblastocytus
      Annulus [Anulus] osseus
        perichondrialis
      Os periosteale reticulofibrosum
      Osteonum primarium
      Osteonum secundarium
Ossificatio endochondrialis
  Centrum ossificationis primarium
    [diaphysiale]
  Gemma osteogenica primaria
  Cavitas medullaris primaria
  Cartilago epiphysialis[33]
    Zona reservata
    Zona proliferativa
      Columella chondrocyti
    Zona hypertrophica
      Chondrocytus hypertrophicus
    Zona resorbens

Cartilago calcificata
Cavitas cartilaginea
Trabecula cartilaginea
Zona ossificationis
  Trabecula ossea primaria
  Trabecula ossea secundaria
  Lamella ossea
  Os endochondriale lamellosum
Centrum ossificationis secundarium
    [epiphysiale]
  Cartilago crescens
  Chondrocytus hypertrophicus
  Cartilago calcificiens
  Lamina ossea
  Gemma osteogenica secundaria
Lacuna erosionis

ARTICULATIONES

Cartilago articularis[33]
  Zona superficialis
  Zona intermedia
  Zona profunda
  Lamina ossea subchondrialis
Capsula articularis
  Stratum synoviale
    Pars villosa
    Pars plana
    Plica synovialis
      Villus synovialis
    Cellula synovialis
      Synoviocytus phagocyticus
      Synoviocytus secretorius
    Lamina propria synovialis
  Stratum fibrosum

LIGAMENTUM (*see* page H 9)

SANGUIS

HEMOCYTI [HAEMOCYTI]

ERYTHROCYTUS
Normocytus

---

[33] From the *Cartilago embryonica* (*Nomina Embryologica*) there develops the bony *Diaphysis* and cartilaginous *Epiphysis*. The cartilage of the epiphysis becomes specialized as the *Cartilago articularis* and *Cartilago epiphysialis*, cartilage of the growth zone at the interphase with the *Diaphysis*. The latter cartilage is commonly called "epiphyseal disc" or "epiphyseal plate" cartilage. As an alternative *Cartilago physealis* has been proposed.

Macrocytus [Megalocytus]
Microcytus
Plasmolemma erythrocyti
Stroma erythrocyti
Erythrocytus polychromatophilicus
Erythrocytus reticulatus
    [Haemoreticulocytus][34]
  Substantia reticularis
  Residuum chromatini [Granulum
    basophilicum]
  Umbra erythrocytica
  Aggregatio erythrocytica[35]

LEUCOCYTUS
Agranulocytus
  Lymphocytus
    Granulum azurophilicum
  Monocytus
    Granulum azurophilicum
Granulocytus
  Granulocytus neutrophilicus
    Granulum neutrophilicum
    Granulum azurophilicum
  Granulocytus neutrophilicus juvenilis
  Granulocytus neutrophilicus
    segmentonuclearis
    Corpusculum chromatini sexualis[36]
  Granulocytus acidophilicus
    [eosinophilicus][37]
    Granulum acidophilicum
    [eosinophilicum][37]
  Granulocytus basophilicus
    Granulum basophilicum

THROMBOCYTUS
Granulomerus
  Granulum thrombocyticum
Hyalomerus

PLASMA SANGUINIS

Hemoconium [Haemo-]
Chylomicronum

PLASMA LYMPHAE

Chylomicronum

HAEMOCYTOPOESIS [HEMO-]

Textus haemopoeticus [hemo-]
Textus myeloideus
  Reticulocytus
  Haemocytoblastus [Hemo-]
Textus lymphoideus
  Reticulocytus
  Lymphoblastus

ERYTHROCYTOPOESIS
Proerythroblastus
Erythroblastus
  Erythroblastus basophilicus
  Erythroblastus polychromatophilicus
  Erythroblastus acidophilicus
  Erythrocytus

GRANULOCYTOPOESIS
Myeloblastus
Promyelocytus
  Promyelocytus neutrophilicus
  Promyelocytus acidophilicus
  Promyelocytus basophilicus
Myelocytus
  Myelocytus neutrophilicus
  Myelocytus acidophilicus
  Myelocytus basophilicus
Metamyelocytus
  Metamyelocytus neutrophilicus
    [Granulocytus neutrophilicus juvenilis]
  Metamyelocytus acidophilicus
    [Granulocytus acidophilicus juvenilis]
  Metamyelocytus basophilicus
    [Granulocytus basophilicus juvenilis]
  Granulocytus neutrophilicus

---

[34] *Reticulocytus* is used for the multipotent, less differentiated stem cell of reticular connecting tissue (*see* footnote 25). The Committee chose *Erythrocytus reticulatus* for the maturing erythrocyte and finds *Hemoreticulocytus* an attractive alternative.

[35] This term signifies a rouleaux formation.

[36] This replaces *Satelles nuclearis* of the first edition *Nomina Histologica*. The *Corpusculum chromatini sexualis* is a satellite of the sex chromosome. Described first by Barr and Bertram (1949, *Nature* 163:676) in nerve cells and by Davidson and Smith (1954, *Br. Med. J.* 2:6) in the mature neutrophilic granulocyte, it has also been called the "Barr body."

[37] *Acido-* replaces the more limited *Eosino-* of the first edition of *Nomina Histologica*. Where possible the Committee avoids dye reference in preference to general staining affinity.

Granulocytus acidophilicus
Granulocytus basophilicus

MEGACARYOCYTOPOESIS
Megacaryoblastus
Megacaryocytus

THROMBOCYTOPOESIS
Megacaryocytus
  Thrombocytus

LYMPHOCYTOPOESIS
Lymphoblastus
Lymphocytus magnus
Lymphocytus medius
Lymphocytus parvus

PLASMOCYTOPOESIS
Plasmoblastus
Plasmocytus

MONOCYTOPOESIS
Monoblastus
Monocytus

TEXTUS MUSCULARIS

TEXTUS MUSCULARIS
  NONSTRIATUS

Myocytus nonstriatus[38]
  Myofilamentum
    Myofilamentum crassum
    Myofilamentum tenue
  Corpusculum insertionis [Area densa]
  Vesicula superficialis
  Nexus

TEXTUS MUSCULARIS STRIATUS
  SKELETALIS

Myocytus skeletalis[38]
  Myofibrilla
  Myofilamentum

Myofilamentum tenue
Myofilamentum crassum
Myomerum
  Discus anisotropicus [Stria A]
  Discus isotropicus [Stria I]
  Zona lucida [Stria H]
  Mesophragma [Linea M]
  Telophragma [Linea Z]
  Reticulum endoplasmaticum
    agranulosum
    Elementum reticulare
    Elementum tubulare
    Cisterna terminalis
  Tubulus transversus
  Trias
Myosatellitocytus
Myocytus rubra
Myocytus alba
Fasciculus muscularis
Endomysium
Perimysium
Epimysium

STRIOMYOHISTOGENESIS

Myoblastus
Myotubus
Myosatellitocytus

TENDO (*see* page H 8)

TEXTUS MUSCULARIS STRIATUS
  CARDIACUS

Myofibra[39]
Myocytus cardiacus[38]
  Discus intercalatus
    Macula adherens
    Nexus
  Myofibrilla
  Reticulum endoplasmaticum
    agranulosum

---

[38] Terms such as "sarcolemma" and "sarcoplasma" or "axolemma" and "axoplasma" which were in the first edition of *Nomina Histologica*, are deleted. The Committee considered such a specific application of the terms *Plasmolemma* and *Cytoplasma* to be tautologous.

[39] A *Myofibra* is a series of cardiac myocytes juxtaposed end-to-end. It is not a synonym for a *Myocytus*.

Tubulus transversus
Dias
Myofibra conducens cardiaca[40]
Myocytus conducens cardiacus[40]

TEXTUS NERVOSUS

NEURONUM [NEUROCYTUS]

Corpus neuroni[38]
  Colliculus axonis
  Neurofilamentum[41]
  Neurofibrilla[41]
  Substantia chromatophilica[42]
Axon [Neuritum]
  Segmentum initiale
  Varicositas axonica
  Telodendron
  Ramus collateralis
  Vesicula synaptica
Dendritum
  Appendix dendritica
    Gemmula dendritica
    Spinula dendritica
Neuronum unipolare
Neuronum pseudounipolare
Neuronum bipolare
Neuronum multipolare
  Neuronum multipolare longiaxonicum
  Neuronum multipolare breviaxonicum
Neuronum secretorium
  Substantia neurosecretoria
Neuronum pigmentosum
Neuromelanocytus

TERMINATIONES NERVORUM

Receptor
  Terminatio nervi libera
  Terminatio folliculi pili
  Corpusculum nervosum terminale
    Corpusculum tactus noncapsulatum
      Torulus tactilis (Car)[43]
      Meniscus tactus

Epithelioidocytus tactus[44]
Corpusculum nervosum capsulatum
  Corpusculum tactus
  Corpusculum lamellosum
    Bulbus internus
    Bulbus externus
    Lamella
  Corpusculum bulboideum
Fusus neurotendineus
Fusus neuromuscularis
  Capsula
    Lamina interna
    Lamina externa
  Myocytus intrafusalis
    Bursa nuclearis myocyti
    Vinculum nucleare myocyti
  Terminatio nervi annulospiralis
  Terminatio nervi racemosa
Effector
  Terminatio neuromuscularis
  Terminatio neuromuscularis fusi
  Plica membranae postsynapticae
  Terminatio neuroepithelialis
    Terminatio neuroglandularis
  Terminatio neurosecretoria
  Terminatio palisadica

SYNAPSIS INTERNEURONALIS

Synapsis vesicularis
  Pars presynaptica
    Vesicula presynaptica
      Vesicula densa
      Vesicula lucida
    Membrana presynaptica
    Densitas presynaptica
  Fissura synaptica
    Substantia intrafissuralis
  Pars postsynaptica
    Membrana postsynaptica
    Densitas postsynaptica
Synapsis nonvesicularis [electricalis]
Synapsis axodendritica
Synapsis axosomatica
Synapsis axoaxonalis

[40] These have been called "Purkinje fibres" and "Purkinje cells" respectively.
[41] *Neurofilamentum* is retained as a general term. With the resolution of light microscopy these may appear aggregated as "Neurofibrilla."
[42] These aggregates of endoplasmic reticulum were formerly called "Nissl substance."
[43] This is a sensory elevation of the skin.
[44] This cell was formerly called a "cell of Merkel."

Synapsis somatodendritica
Synapsis somatosomatica
Synapsis dendrodendritica
Bulbus terminalis
Bulbus preterminalis
Pes terminalis
Calix terminalis
Synapsis invaginata

NEUROGLIA

Gliocytus
    Corpus gliocyti
    Processus gliocyti
    Gliofilamentum[45]
Gliocytus centralis
    Ependymocytus
        Ependymocytus columnaris
        Ependymocytus ciliatus
        Ependymocytus choroideus
        Ependymocytus taeniatus [ten-]
    Astrocytus
        Astrocytus protoplasmaticus
        Astrocytus fibrosus
            Processus vascularis
            Processus pialis
            Membrana limitans gliae
                superficialis
            Membrana limitans gliae
                perivascularis
            Membrana limitans gliae
                periventricularis
    Oligodendrocytus
        Processus myelinopoeticus
    Microglia
Gliocytus periphericus
    Gliocytus ganglii[46]
    Neurolemmocytus

Gliocytus terminalis
Neuropilus[47]

INVESTIO PROCESSUS NEURONI

Oligodendrocytus
Neurolemma[48]
    Neurolemmocytus
    Mesaxon
Neurofibra myelinata
    Stratum myelini
    Segmentum internodale
        Lamella myelini
        Manica lamellaris terminalis
    Nodus neurofibrae
    Incisura myelini
Neurofibra nonmyelinata

HISTOLOGIA SPECIALIS

SPLANCHNOLOGIA

NOMINA GENERALIA

Tunica mucosa[49]
    Epithelium mucosae
    Lamina propria mucosae
    Lamina muscularis mucosae
Tela submucosa
Tunica muscularis
    Stratum circulare
    Stratum longitudinale
Tunica adventitia [Tunica fibrosa, Capsula]
Tela subserosa
Tunica serosa
    Lamina propria serosae
    Mesothelium
    Pericardium (*see* page H 28)

---

[45] *Gliofilamentum* is retained as a general term. With the resolution of light microscopy these may appear aggregated as "gliofibrilla."

[46] The *Gliocytus ganglii* is a *Neurolemmocytus* adjacent to the *Corpus neuronum* of most neurons in the PARS PERIPHERICA. These have also been called "neurosatellite cells" or "satellite cells."

[47] The *Neuropilus* is the meshwork formed of axons, dendrites, and gliocyte processes.

[48] The *Neurolemma* includes the outermost part of the *Neurolemmocytus* cytoplasm and its glycocalyx.

[49] Some authors distinguish between two types of *Tunicae mucosae*. A mucous membrane having numerous glands near the surface and simple or pseudostratified secretory epithelium is called *Tunica mucosa glandularis*; a mucous membrane having glands removed from the surface and a stratified squamous nonsecretory epithelium is called *Tunica mucosa nonglandularis (cutanea)*. Examples of organs having the former lining tunica are the *Gaster* and *Uterus*; examples of the latter are the *Oesophagus* and *Vestibule*.

Peritoneum (*see* page H 26)
Pleura (*see* page H 20)
Vas sanguineum
  Rete capillare subepitheliale
  Plexus vascularis
    Rete arteriale
    Plexus venosus
  Plexus vascularis submucosus
    Rete arteriale submucosum
    Plexus venosus submucosus
  Plexus vascularis intramuscularis
    Rete arteriale intramusculare
    Plexus venosus intramuscularis
Vas lymphaticum intrinsecum
  Rete lymphocapillare subepitheliale
  Plexus lymphaticus intramucosus
  Plexus lymphaticus submucosus
  Plexus lymphaticus subserosus
Plexus nervorum intrinsecus (intramuralis)
  Plexus nervorum submucosus
  Plexus nervorum myentericus
  Ganglion plexus
Parenchyma
Stroma
Glandula

APPARATUS DIGESTORIUS
[SYSTEMA DIGESTORIUM]

CAVITAS ORIS
Bucca
Labium
  Pars cutanea
  Pars intermedia
  Pars mucosa
Palatum
  Palatum durum
    Tunica mucosa
      Zona adiposa
      Zona glandularis
      Zona fibrosa
    Stratum cavernosum (Car, Un)[50]
  Palatum molle [Velum palatinum]
    Facies nasopharyngea
      Tunica mucosa respiratoria
    Facies oropharyngea

    Tunica mucosa oralis
    Stratum elasticum
    Tela submucosa
      Glandula palatina
    Lamina tendinomuscularis
Gingiva
  Pars libera
  Pars fixa
  Margo gingivalis
  Papilla gingivalis (interdentalis)
  Sulcus gingivalis
  Fibra gingivalis

Glandula salivaria (*see* GLANDULA UT
    ORGANUM page H 7)
Portio terminalis
  Acinus
  Alveolus
  Tubulus
  Tubuloacinus
  Tubuloalveolus
  Mucocytus
  Serocytus
  Semiluna serosa
Ductus intercalatus
Ductus striatus
Ductus intralobularis
Ductus interlobularis
Ductus interlobaris
Ductus excretorius [glandulae]
Glandula labialis
Glandula palatina

Dens
Corona dentis
Cervix dentis
Radix [Radices] dentis
Cavum pulpare dentis
  Cavum coronale dentis
  Canalis radicis dentis
    Foramen apicale dentis
Enamelum
Enameloblastus [Ameloblastus][51]
  Prisma enameli
    Membrana prismatis
    Crystallum hydroxyapatiti

---

[50] This is a layer of cavernous tissue in the *Tela submucosa* of the hard palate, consisting of a dense plexus of muscular veins with many arteriovenous anastomoses. It is best developed in the horse.
[51] In ungulates the enameloblasts persist in certain teeth long after the usual odontogenesis.

Cuticula dentis
Lamella enameli
Fusus enameli
Fasciculus enameli
Junctio dentino-enameli
Linea incrementalis enameli
   Linea neonatalis
Dentinum
   Tubulus [Canaliculus] dentinalis
      Dentinum peritubulare
      Ramus lateralis
   Processus dentinoblasti
   Lamella dentinalis
   Substantia fundamentalis
      Crystallum hydroxyapatiti
   Linea incrementalis dentini
   Stratum granulosum dentini radicis
   Globulus dentinalis
   Spatium interglobulare
   Dentinum juxtapulpare
   Predentinum
      Fibra predentinalis
   Dentinum secundarium
   Fibra dentini
   Junctura dentinocementi
Cementum
   Cementum noncellulare
   Cementum cellulare
   Cementocytus
   Substantia intercellularis
      Lacuna
      Canaliculus
      Substantia fundamentalis
      Fibra cementalis
      Crystallum hydroxyapatiti
   Fibra perforans cementalis
Pulpa dentis
   Cornu pulpae
   Pulpocytus [Reticulocytus]
   Dentinoblastus [Odontoblastus]
   Stratum subdentinoblasticum
   Plexus nervorum subdentinoblasticus
   Pulpa radicularis
Periodontium[52]
   Ligamentum periodontale

Ligamentum gingivale
Ligamentum dentoalveolare
   Fibra cementoalveolaris
   Fibra interradicularis
   Fibra apicalis
   Fibra interdentalis
Periosteum alveolare
Alveolus dentalis
Septum interradiculare
Septum interalveolare

Odontogenesis
Lamina dentalis
   Lamina dentis ingualis[53]
   Lamina dentis vestibularis[53]
   Gemma dentis
Germen dentis
   Organum enameleum
      Epithelium enameleum externum
      Pulpa enamelea [Reticulum stellatum]
      Stratum intermedium
      Epithelium enameleum internum
         Enameloblastus [Ameloblastus][54]
      Cuticula enameli
      Cuspis dentis
      Residuum epitheliale
      Papilla dentis
      Sacculus dentis

Lingua
Papilla lingualis
Papilla filiformis
Papilla conica (Ru)
Papilla lentiformis
Papilla fungiformis
Papilla foliata
   Folium papillae
   Sulcus papillae
Papilla vallata
   Vallum papillae
   Sulcus papillae
Papilla marginalis (Car, su)[55]
Glandula ingualis

[52] In clinical parlance periodontists include not only the *Ligamentum periodontale* but also the *Periosteum alveolare* as *Periodontium*. The various *Fibrae* were formerly called "Sharpey's fibers."
[53] These terms replace *Lamina dentis medialis* and *Lamina dentis lateralis*, respectively, of the first edition of *Nomina Histologica*.
[54] In ungulates the enameloblasts persist in certain teeth long after the usual odontogenesis.
[55] These papillae are present in newborn carnivora and swine.

Glandula lingualis apicalis[56]
Glandula radicis linguae[56]
Glandula gustatoria
Tonsilla lingualis
   Crypta
   Papilla tonsillaris (su)
Gemma gustatoria (*see* page 36)

PHARYNX

Fauces
Tonsilla palatina
   Folliculus tonsillaris
      Fossula tonsillaris
         Crypta tonsillaris
      Nodulus lymphaticus
   Capsula tonsillaris
   Sinus tonsillaris (Ru)[57]
Tonsilla veli palatini (Car, Un)
Tonsilla para-epiglottica (fe, su, ov, cap)
   Sulcus tonsillaris (su)
Glandula pharyngealis
Tonsilla pharyngealis [adenoidea][58]
Tonsilla tubaria

OESOPHAGUS [ESOPHAGUS]
Glandula oesophagea propria
   Pars terminalis tubuloalveolaris
      Mucocytus
   Ampulla ductus
Glandula cardiaca oesophagi
   Pars terminalis tubularis

GASTER [VENTRICULUS]
Foveola gastrica
Epitheliocytus superficialis gastricus

Glandula cardiaca
   Exocrinocytus cardiacus
Glandula gastrica propria
   Isthmus
      Epitheliocytus nondifferentiatus
   Cervix
      Mucocytus cervicalis
   Pars principalis [corpus et fundus]
      Exocrinocytus principalis
         Granulum zymogeni
      Exocrinocytus parietalis
         Canaliculus intracellularis
Glandula pylorica
   Exocrinocytus pyloricus
Endocrinocytus gastrointestinalis[59]
Stratum compactum (Car)[60]
Nodulus lymphaticus
Pars nonglandularis (Un)[61]
   Papilla ruminis (Ru)
   Cellula reticuli (Ru)
   Crista reticuli (Ru)
   Papilla reticuli (Ru)
Papilla unguiculiformis (Ru)
   Lamina omasi (Ru)
   Papilla omasi (Ru)
   Epithelium stratificatum squamosum
      cornificatum
   Lamina propria mucosae
      Stratum papillare
         Micropapilla
      Stratum reticulare

INTESTINUM TENUE
Plicae circulares
Villus intestinalis
   Epitheliocytus columnaris villi
      Limbus striatus

---

[56] These terms replace *Glandula anterior* and *Glandula posterior*, respectively, of the first edition of *Nomina Histologica*.

[57] The *Nomina Anatomica* has replaced *Sinus tonsillaris* with the more accurate *Fossa tonsillaris*. The term Sinus is now used to designate the deep, narrow-mouthed cavity in the palatine tonsil of Ruminantia. A *Folliculus tonsillaris* is a *Fossula*, ending as *cryptae*, with its surrounding lymphatic tissue. Folliculus was included in the first edition of *Nomina Histologica* as an alternative for *Nodulus*. In this context "folliculus" is a misnomer, based upon the misconception that nodules are cavitated.

[58] Changed in this edition from *nasopharyngealis* to conform with the *Nomina Anatomica* (page A 34).

[59] This term replaced *Endocrinocytus gastricus*. These endocrine cells are found in the *Gaster, Intestinum tenue et crassum*, and *Pancreas* as well as in the lining epithelium of the respiratory apparatus. These have been called "argentaffin," "argyrophilic," or "chromaffin," depending upon dye affinities of their cytoplasmic granules. Their metabolic conversion of amines, i.e. amine precursor uptake and decarboxylation, has resulted in the acronym "APUD cells."

[60] The *Stratum compactum* is a layer of dense collagenous connective tissue between the base of the mucosal glands and the *Lamina muscularis mucosae*.

[61] The term designates the part of the stomach that is lined by stratified squamous epithelium in ungulates.

H 17

Microvillus
Exocrinocytus caliciformis
Stroma villi
Myocytus villi
Vas lymphaticum centrale
Glandula [Crypta] intestinalis
Epitheliocytus nondifferentiatus
Exocrinocytus columnaris
Exocrinocytus caliciformis
Exocrinocytus cum granulis acidophilis[62]
Endocrinocytus gastrointestinalis[59]
Glandula submucosae
Stratum compactum (Car)[60]
Nodulus lymphaticus solitarius
Noduli lymphatici aggregati

INTESTINUM CRASSUM
Appendix vermiformis
Epitheliocytus superficialis intestinalis
Limbus striatus
Microvillus
Glandula [Crypta] intestinalis
Epitheliocytus nondifferentiatus
Epitheliocytus columnaris
Exocrinocytus caliciformis
Endocrinocytus gastrointestinalis[59]
Stratum compactum (Car)[60]
Nodulus lymphaticus solitarius
Noduli lymphatici aggregati (Car, Un)

Canalis analis
Glandula analis
Linea anorectalis
Linea anocutanea
Sinus paranalis (Car)
Glandula sinus paranalis
Glandula apocrina
Glandula sebacea (fe)
Glandula circumanalis (ca)[63]

HEPAR
Tunica serosa
Tela subserosa
Tunica fibrosa
Lobulus hepaticus
Zona centralis
Zona intermedia
Zona peripheralis
Lamina hepatica
Epitheliocytus hepatis [Hepatocytus]
Lamina hepatica limitans
Canalis portalis
Trias hepatica
Arteria interlobularis
Vena interlobularis
Vas capillare interlobulare
Vas sinusoideum
Macrophagocytus stellatus[64]
Spatium perisinusoideum
Lipocytus perisinusoideus
Vena centralis
Vena sublobularis
Vena occludens (ca)
Canaliculus bilifer
Ductulus bilifer
Ductus interlobularis bilifer
Capsula fibrosa perivascularis[65]

Vesica biliaris [fellea]
Fundus, Corpus
Tunica mucosa
Epitheliocytus superficialis
Plica tunicae mucosae
Crypta tunicae mucosae
Glandula tunicae mucosae (Car, Un)
Exocrinocytus mucosus [Mucocytus]
Exocrinocytus caliciformis
Cervix [Collum]
Glandula mucosa

---

[62] This cell was called *Cellula panethensis* in the first edition of *Nomina Histologica.* The Committee dropped the term *panethiensis*, an adjectival construct from an eponym, in deference to the principle rejecting eponyms.

[63] In the adult dog the main mass of the ring of glands in a cutaneous zone of the anal canal consists of lobules of large polygonal cells. They are derived from sebaceous glands but have no ducts.

[64] *Reticuloendotheliocytus* in the first edition of *Nomina Histologica* was replaced by *Macrophagocytus.* These cells of the hepatic sinusoid were formerly called "Kupffer cells." Macrophages of various locations have, as a consequence of their similar structure and function, been grouped and called "macrophage system," formerly the "reticuloendothelial system" (*see* footnote 34).

[65] This term, adopted from the *Nomina Anatomica*, is the connecting tissue sheath surrounding the portal vein, hepatic artery, and common bile duct at the porta, continuing into the interior of the liver with the branches of these vessels.

Ductus cysticus
  Plica spiralis
Ductus hepatocysticus

Ductus choledochus[66]
Tunica mucosa
  Exocrinocytus caliciformis
  Glandula tunicae mucosae
    Exocrinocytus mucosus [Mucocytus]
Tunica muscularis
  Musculus sphincter ductus choledochi
Tunica adventitia
Ampulla hepatopancreatica
  Musculus sphincter ampullae

PANCREAS
Pars exocrina pancreatis
  Lobulus pancreaticus
  Septum interlobulare
  Acinus pancreaticus
    Exocrinocytus pancreaticus
      [Acinocytus]
      Granulum zymogeni
    Epitheliocytus centroacinosus
Ductus intercalatus
Ductus intralobularis
Ductus interlobularis
Pars endocrina pancreatis (*see* INSULAE
  PANCREATICAE page H 27)

APPARATUS RESPIRATORIUS
[SYSTEMA RESPIRATORIUM]

CAVITAS NASI
Regio cutanea
  Vestibulum nasi
    Vibrissae
Regio respiratoria
  Tunica mucosa respiratoria
    Epithelium columnare
      pseudostratificatum ciliatum
      Epitheliocytus ciliatus
      Exocrinocytus caliciformis
      Epitheliocytus microvillosus
      Epitheliocytus basalis
    Glandula nasalis
    Glandula nasalis lateralis (Car, su, ov,
      cap, eq)
  Stratum cavernosum
    Plexus cavernosus concharum

Regio olfactoria
Tunica mucosa olfactoria (*see* ORGANUM
  OLFACTUS, page H 32)

CAVITAS LARYNGIS
Vestibulum laryngis
  Epithelium stratificatum squamosum
    noncornificatum
  Epithelium pseudostratificatum
    columnare ciliatum
  Glandula epiglottica
  Glandula arytenoidea
  Gemma gustatoria epiglottidis
  Plica vestibularis
  Ventriculus laryngis
  Epithelium pseudostratificatum
    columnare ciliatum
  Glandula ventriculi laryngis
  Sacculus laryngis (Homo)
    Glandula sacculi laryngis
Glottis
  Plica vocalis
  Pars intermembranacea
  Epithelium stratificatum squamosum
    noncornificatum
Cavitas infraglottica
  Epithelium pseudostratificatum
    columnare ciliatum
  Glandula laryngea
Nodulus lymphaticus solitarius
Noduli lymphatici aggregati

TRACHEA ET BRONCHI
Tunica mucosa respiratoria
  Epithelium pseudostratificatum
    columnare ciliatum
    Epitheliocytus ciliatus
    Epitheliocytus microvillosus
    Exocrinocytus caliciformis
  Lamina propria mucosae
    Lamina fibrarum elasticarum
Tela submucosa
Glandulae tracheales et bronchiales
Noduli lymphatici tracheales et bronchiales
Tunica fibromusculocartilaginea
  Ligamenta annularia [anularia]
  Musculus trachealis et bronchialis
  Cartilagines tracheales et bronchiales
Paries membranaceus
Tunica adventitia

---

[66] Terms under this heading were added in this edition.

PULMO

Arbor bronchialis

Bronchus

Tunica mucosa respiratoria

Epithelium pseudostratificatum
ciliatum[67]

Lamina propria mucosae

Lamina fibrarum elasticarum

Lamina muscularis mucosae

(Tela submucosa)

Tunica musculocartilaginea

Musculus spiralis

Cartilagines bronchiales

Glandula bronchialis

Tunica adventitia

Bronchiolus

Tunica mucosa respiratoria

Epithelium pseudostratificatum
ciliatum[67]

Epithelium simplex ciliatrum[67]

Exocrinocytus caliciformis

Lamina propria mucosae

Fibra elastica longitudinalis

Lamina muscularis mucosae

(Tela submucosa)

Tunica muscularis

Tunica adventitia

Lobulus pulmonis secundarius[68]

Bronchiolus terminalis

Epithelium simplex cuboideum ciliatum

Exocrinocytus bronchiolaris[69]

Arbor alveolaris [Acinus pulmonaris][70]

Lobulus pulmonis primarius[68]

Bronchiolus respiratorius

Epithelium simplex cuboideum
ciliatum

Exocrinocytus bronchiolaris[69]

Fibra elastica longitudinalis

Musculus spiralis

Ductus alveolaris

Epithelium simplex cuboideum
ciliatum

Fibra elastica

Myocytus nonstriatus

Atrium alveolare

Sacculus alveolaris

Alveolus pulmonis

Epithelium simplex squamosum

Epitheliocytus respiratorius

Epitheliocytusmagnus[granularis][71]

Macrophagocytus alveolaris

Macrophagocytus pulvereus

Septum interalveolare

Porus septi

PLEURA

Tunica serosa

Mesothelium

Lamina propria

Tela subserosa

Mediastinum

Fenestra mediastinalis

APPARATUS UROGENITALIS

ORGANA URINARIA

Ren

Capsula adiposa

Capsula fibrosa

Lobus renalis

Cortex renalis

Zona externa [peripherica]

Zona interna [juxtamedullaris]

Lobulus corticalis

Pars convoluta

Pars radiata

Medulla renalis

Zona externa

Zona interna

---

[67] The epithelium is pseudostratified columnar in the primary branches and distally decreases in height to pseudostratified cuboidal. Likewise, simple columnar epithelium becomes simple cuboidal in the terminal bronchiole.

[68] A secondary pulmonary lobule consists of about 50 primary lobules and is delineated by interlobular septa. A primary pulmonary lobule consists of a respiratory bronchiole together with its associated alveolar ducts, alveolar sacs, and alveoli.

[69] The nonciliated cells occurring in the epithelium of the bronchioles are presumably secretory and were formerly called "Clara cells."

[70] When the pulmonary alveolar tree is reconstructed, the appearance is that of a berry or drupe (*acinus*) composed of drupelets, the alveolar sacs.

[71] This alveolar cell is also called a "type II cell," "granular pneumocyte," and "septal cell." It contains lamellar bodies or cytosomes and is believed to be the source of surfactant.

Pyramis renalis
  Basis pyramidis
  Papilla renalis
    Crista renalis (Car, ov, cap, eq)
    Area cribrosa
      Foramen papillare
Columna renalis
Corpusculum renale[72]
  Polus vascularis
  Glomerulus
    Rete capillare glomerulare
    Vas capillare glomerulare
      Endotheliocytus fenestratus
        Membrana basalis
    Mesangium
      Lamella hyalina
      Mesangiocytus
    Capsula glomeruli (*see* below)
  Polus tubularis
Tubulus renalis
Nephronum[73]
  Nephronum breva [corticale]
  Nephronum intermedium
  Nephronum longum [juxtamedullare]
  Capsula glomeruli
    Paries externa
    Paries interna
    Podocytus
      Cytotrabecula
      Cytopodium
    Lumen capsulae
  Tubulus contortus proximalis[74]
  Tubulus rectus proximalis[74]
    Epithelium simplex cuboideum
      Epitheliocytus microvillosus
        Limbus penicillatus
        Limbus striatus basalis
  Tubulus attenuatus[74]
    Pars descendens
    Pars ascendens
    Epithelium simplex squamosum
  Tubulus rectus distalis[74]
    Epithelium simplex cuboideum

    Macula densa
  Ansa nephroni[74]
  Tubulus contortus distalis[74]
    Epithelium simplex cuboideum
Tubulus renalis colligens
  Tubulus renalis arcuatus
  Tubulus colligens rectus
  Ductus papillaris
    Epithelium simplex cuboideum

Vasa sanguinea renalia
Arteria interlobaris
Arteria arcuata
Arteria interlobularis
Arteria intralobularis
Arteriola glomerularis afferens
Rete capillare glomerulare
Arteriola glomerularis efferens
Rete capillare peritubulare corticalis et
      medullaris
Fasciculus vascularis [Vasa recta]
Vena interlobaris
Vena arcuata
  Vena interlobularis
    Vena intralobularis
      Venula stellata
  Venula recta
Vena capsularis (fe)

Complexus juxtaglomerularis
Macula densa
  Epitheliocytus maculae densae
Tunica media arteriolae glomerularis
  Endocrinocytus myoideus
  [Juxtaglomerulocytus]
Insula perivascularis mesangii
  Mesangiocytus

Pelvis renalis
Tunica mucosa
  Epithelium transitionale
  Glandula pelvis renalis (eq)
    Exocrinocytus mucosus

---

[72] A *Corpusculum renale* is a composite of the *Glomerulus* and the *Capsula glomeruli*, the latter being the most proximal part of the *Nephronum*.

[73] A *Nephronum* is composed of continuous parts, all derived from the nephrogenic vesicle, beginning proximally with the *Capsula glomeruli* and continuing to, but not including, the *Tubulus renalis colligens*.

[74] *Tubulus contortus proximalis* and *Tubulus rectus proximalis* replace *Pars proximalis tubuli nephroni*, *Partes convoluta et recta*; *Tubulus rectus distalis* and *Tubulus contortus distalis* replace *Pars distalis tubuli nephroni (Pars recta, Pars convoluta,* and *Pars conjungens),* the replaced terms being of the first edition of *Nomina Histologica.* The *Tubulus rectus proximalis, Tubulus attenuatus,* and *Tubulus rectus distalis* constitute a segment of the *Nephronum* known as the *Ansa nephroni.*

Tunica muscularis
  Stratum longitudinale
  Stratum circulare
Tunica adventitia

Ureter
Tunica mucosa
  Epithelium transitionale
  Glandula ureterica (eq)
Tunica muscularis
  Stratum longitudinale internum
  Stratum circulare
  Stratum longitudinale externum
Tunica adventita

Vesica urinaria
Tunica mucosa
  Epithelium transitionale
  Glandula trigoni vesicae
Tunica muscularis
  Stratum longitudinale internum
  Stratum circulare
  Stratum longitudinale externum
Tunica serosa
  Tela subserosa
Tunica adventitia

ORGANA GENITALIA MASCULINA

Testis
Tunica vaginalis
  Lamina parietalis
  Lamina visceralis
Tunica albuginea
Tunica vasculosa
Mediastinum testis
Septulum testis
Lobulus testis
Interstitum testis
  Endocrinocytus interstitialis
    Crystalloidum
Parenchyma testis
  Tubulus seminifer convolutus
  Epithelium spermatogenicum[75]
    Epitheliocytus sustentans

Cellulae spermatogenicae
Lamina limitans
  Membrana basalis
  Stratum myoideum
  Stratum fibrosum
Tubulus seminifer rectus
Rete testis
Ductulus efferens (*see Epididymis*, page H 23)

Spermatogenesis
Spermatogonium
  Spermatogonium A
  Spermatogonium intermedium
  Spermatogonium B
Spermatocytus primarius
Spermatocytus secundarius
Spermatidium[76]
  Idiosoma
    Diplosoma
    Complexus golgiensis
  Proacrosoma
    Vesicula proacrosomatica
    Granulum proacrosomaticum
  Acrosoma [Galea acrosomatica]
    Membrana acrosomatica externa
    Membrana acrosomatica interna
    Substantia acrosomatica
  Corpusculum chromatoideum
  Flagellum
    Annulus [Anulus]
  Vagina cytoplasmatica
  Plicatura equatorialis
  Annulus [Anulus] equatorialis
  Manicula caudalis
    Microtubulus
  Corpus residuale
Spermatozoön [Spermium]
  Caput
    Nucleus
    Nucleus
      Vesicula nuclearis
      Sacculus nuclearis
    Acrosoma [Galea acrosomatica]
      Membrana acrosomatica interna

---

[75] This epithelium is not classified according to criteria applied to the somatic types (see pages H 6 and H 7). The *Cellulae spermatogenicae* reside in an epithelium of sustentacular cells, which, when the germ cells are not proliferated, e.g. autoimmune aspermatogenesis and abdominal cryptorchidism, is for the most part a simple columnar epithelium. This distinction, based upon structure, parallels the concept of Weismann (1893, *The Germ-plasm*, Scott, London) of "germ" and "soma."

[76] These terms descriptive of "spermiogenesis," that portion of *Spermatogenesis* during which the *Spermatidium* is converted to a *Spermatozoon*, are new to this edition.

Membrana acrosomatica externa
Substantia acrosomatica
Granulum acrosomale
Perforatorium
Substantia subacrosomatica
Substantia postacrosomatica
Annulus [Anulus] caudalis
Fossula articularis
Flagellum
Pars conjungens
Patella basalis
Basis columnaris
Capitulum
Columna striata
Centriolum proximale
Centriolum distale
Pars intermedia
Axonema [Filamentum axiale]
Microtubulus centralis
Diplomicrotubulus periphericus
Fibra densa
Vagina mitochondrialis
Annulus [Anulus]
Pars principalis
Axonema
Fibra densa
Vagina fibrosa
Columna longitudinalis
Costa fibrosa
Pars terminalis
Axonema

Epididymis
Ductulus efferens testis[77]
Epithelium pseudostratificatum
columnare
Stratum fibromusculare
Septum epididymidis
Lobulus epididymidis
Ductus epididymidis
Epithelium pseudostratificatum
columnare
Epitheliocytus microvillosus[78]
Limbus penicillatus
Epitheliocytus basalis

Tunica fibromuscularis
Tunica adventitia
Tunica serosa
Tela subserosa
(Appendix testis)
(Ductulus aberrans)
(Appendix epididymidis)
(Paradidymidis)
(Ductulus paradidymidis)

Ductus deferens
Tunica mucosa
Plica mucosae
Epithelium pseudostratificatum
columnare
Epitheliocytus microvillosus
Limbus penicillatus
Epitheliocytus basalis
Tunica muscularis
Stratum longitudinale internum
Stratum circulare
Stratum longitudinale externum
Tunica adventitia
Tunica serosa
Tela subserosa
Ampulla ductus deferentis
Diverticulum ampullae
Glandula ampullae (Car, Un)
Ductus ejaculatorius

GLANDULAE GENITALES ACCESSORIAE

Ampulla ductus deferentis

Vesicula [Glandula] seminalis (Homo, eq)
[Glandula vesicularis (su, Ru)]
Lobulus glandularis (su, Ru)
Portio terminalis
Alveolus
Exocrinocytus mucosus
Tunica muscularis
Tunica adventitia
Tunica serosa
Tela subserosa

---

[77] In deference to their embryologic origin from the *Tubuli mesonephrici (Nomina Embryologica* page E 18) and their forming a part of the *Caput epididymidis,* the *Ductuli efferentes* are included in this edition as part of the *Epididymis.*
[78] Microvilli of these cells are tall and branched. These have been called "stereocilia" (see footnote 2).

Prostata
Pars disseminata prostatae[79]
Corpus prostatae[79]
Capsula prostatae
   Stratum fibrosum
   Stratum musculare
Stroma myoelasticum
   Septum prostaticum
Parenchyma
   Lobus
      Portio terminalis
         Tubuloalveolus
      Exocrinocytus mucosus
      Concretio prostatica
   Ductulus prostaticus
   Ductus prostaticus

Glandula bulbourethralis (Homo, fe, Un)
Portio terminalis
   Tubuloacinus
   Tubuloalveolus
      Exocrinocytus mucosus
Ductus glandulae bulbourethralis
Tunica adventitia

Penis
Cutis penis
Preputium
   Glandula sebacea [preputialis]
      Epitheliocytus sebaceus [Sebocytus]
Cutis glandis
   Spina glandis (fe)
Fascia penis superficialis
   Tunica dartos
Fascia penis profunda
Corpus cavernosum
   Tunica albuginea
   Trabecula
   Caverna
   Arteria helicina
   Vena cavernosa
   Vena emissaria
   Os penis (Car)

Corpus spongiosum
   Tunica albuginea
   Trabecula
   Caverna

Urethra masculina
Pars prostatica
   Epithelium transitionale
      Exocrinocytus mucosus
   Stratum spongiosum (Car, Un)
   Glandula collicularis
   Utriculus prostaticus [Uterus masculinus]
Pars membranacea
   Epithelium pseudostratificatum
      columnare
   Tunica muscularis
      Stratum longitudinale
      Stratum circulare
Pars spongiosa
   Epithelium pseudostratificatum
      columnare
   Fossa navicularis (urethrae)
      Epithelium squamosum stratificatium
Lacunae urethrales
Glandula urethralis[79]
Ductus [Canalis] paraurethralis[80]
Nodulus lymphaticus

Scrotum[81]
Cutis scroti
Tunica dartos

ORGANA GENITALIA FEMININA

Ovarium
Epithelium superficiale
   Mesotheliocytus cuboideus microvillosus
Tunica albuginea
Stroma ovarii
   Interstitium ovarii
      Textus connectivus cellularis
      Endocrinocytus interstitialis
Cortex ovarii [Zona parenchymatosa]
   Fossa ovarii (eq)[82]

---

[79] In Carnivora the prostate is most like that of man with distinct lateral lobes. The term *Corpus prostatae* is also applicable to the boar, bull and stallion, but not the ram and billy goat. A *Pars disseminata* is well developed in the pig ruminants. The urethral glands were formerly called "glands of Littre."

[80] This term added in accordance with the *Nomina Anatomica.*

[81] These terms added in this edition.

[82] This is a notch-like depression of the surface of the mare's ovary, also called the "ovulation fossa." It is not to be confused with the *Fossa ovarica* of the *Nomina Anatomica* (page A 46). The latter is a shallow depression in the lateral wall of the pelvis containing the ovary in the mulliparous woman.

Folliculus ovaricus primordialis[83]
Folliculus ovaricus primarius[83]
Folliculus ovaricus secundarius[83]
Folliculus ovaricus tertiarius[83]
    [vesiculosus][83]
  Theca externa
  Theca interna
    Endocrinocytus thecalis
  Membrana basalis
  Epithelium folliculare [Stratum
    granulosum]
    Epitheliocytus follicularis
  Antrum folliculare
  Liquor follicularis
  Cumulus oöphorus
  Corona radiata
    Zona pellucida
    Ovocytus
Stigma folliculare
Folliculus atreticus
Corpus atreticum
Corpus hemorrhagicum[84]
Corpus luteum
  Corpus luteum cyclicum
    [menstruationis]
  Corpus luteum graviditatis
    Endocrinocytus corporis lutei
      [Luteocytus]
      Granulosoluteocytus
      Thecaluteocytus
  Corpus luteum regressum[84]
Corpus albicans
Medulla ovarii [Zona vasculosa]
  Chorda medullaris
  Tubulus medullaris
  Endocrinocytus interstitialis
  Rete ovarii

Ovogenesis
Ovogonium
Ovocytus primarius
Ovocytus secundarius
Polocytus primarius
Polocytus secundarius

Ovum

Epoöphoron[84]
Ductus epoöphorontis longitudinalis
  Ductulus transversus
Appendix vesiculosus

Paroöphoron[84]
Ductulus paroöphorontis

Tuba uterina [Salpinx]
Tunica mucosa
  Plica
  Epithelium simplex columnare
    Epitheliocytus ciliatus
    Epitheliocytus microvillosus
  Lamina propria mucosae
    Textus connectivus cellulosus
Tunica muscularis
  Stratum circulare
  Stratum longitudinale
Tela subserosa
Tunica serosa

Uterus
Tunica mucosa [Endometrium]
  Epithelium simplex columnare
    Epitheliocytus ciliatus
    Epitheliocytus microvillosus
  Epithelium pseudostratificatum
    columnare (su, Ru)
  Lamina propria mucosae [Stroma
    endometrialis]
  Stratum functionale endometrii
    Textus connectivus cellulosus
      Deciduocytus
    Stratum compactum endometrii
    Stratum spongiosum endometrii
    Glandula uterina
  Stratum basale endometrii
  Arteria spiralis
  Arteria basalis
Caruncula (Ru)
Microcaruncula (eq)[84]

---

[83] A primordial follicle is a small, primary ovocyte enveloped by a single layer of flattened follicular epitheliocytes; a primary follicle is a primary ovocyte enveloped by a single layer of cuboidal or columnar follicular epitheliocytes; a secondary follicle is a growing primary ovocyte enveloped by a stratified follicular epithelium and a developing follicular theca; a teritiary follicle is a large primary ovocyte enveloped by an antrum-containing, follicular epithelium and a well-developed theca. Tertiary follicles are also called "vesicular follicles"; preovulatory follicles are called "mature follicles" or "Graafian follicles."

[84] These terms were added in this edition.

Tunica muscularis [Myometrium]
  Stratum submucosum
  Stratum vasculosum
  Stratum supravasculosum
  Stratum subserosum
Cornu uterinum (Car, Un)
  Stratum musculare circulare
  Stratum vasculare
  Stratum musculare longitudinale
Tela subserosa
Tunica serosa [Perimetrium]
Tunica adventitia
Parametrium
Cervix uteri
  Portio prevaginalis (cervicis)[84]
    Epithelium simplex columnare
      Epitheliocytus superficialis
        Exocrinocytus mucosus
      Epitheliocytus ciliatus
    Glandula cervicalis uteri
      Exocrinocytus mucosus
    Vesicula cervicalis
Plicae palmatae (Homo)
Plica circularis (bo)
Pulvinus cervicalis (su, ov, cap)
Portio vaginalis (cervicis)
  Epithelium stratificatum squamosum

Vagina
Tunica mucosa
  Epithelium stratificatum squamosum
  Epithelium stratificatum squamosum
    cornescens (Car, su, ov)[85]
  Epithelium stratificatum squamosum
    non-cornificatum (bo, eq)
    Exocrinocytus mucosus
  Lamina propria mucosae
Tunica muscularis
  Fasciculus circularis
  Stratum longitudinale
Tunica adventitia

Vulva [Pudendum femininum][86]
Labium majus pudendi
Labium minus pudendi
Vestibulum vaginae

Epithelium stratificatum squamosum
Lamina propria mucosae
Glandula vestibularis minor
Glandula vestibularis major
  Lobulus glandularis
    Portio terminalis alveolaris
      Exocrinocytus mucosus
    Portio terminalis tubulosa
    Ductus interlobularis
      Sinus ductus
    Ductus glandulae
Bulbus vestibuli
Clitoris
  Corpora cavernosa clitoridis
    Tunica albuginea
    Trabeculae
    Cavernae
  Corpus spongiosum
    Tunica albuginea
    Trabeculae
    Cavernae

Urethra feminina
Tunica mucosa
  Epithelium pseudostratificatum
  Epithelium stratificatum squamosum
  Glandula urethralis
  Lacunae urethrales
Stratum spongiosum
Tunica muscularis
(Ductus paraurethralis)[87]
(Glandulae paraurethrales)[87]

PERITONEUM

Tunica serosa
  Mesothelium
  Lamina propria
Tela subserosa
Omentum
Tunica serosa
Tela subserosa
Trabecula omentalis
Lobulus adiposus
Macula lactea
Fenestra omentalis

[85] This epithelium is cornified periodically. This cornifying epithelium is different from the continuously cornified epithelium found elsewhere.

[86] The human vestibule is shallow and part of the *Pudendum feminium.* In quadrupeds it is relatively longer, and not external organ, and therefore it is not included in *Pudendum feminium* in the *Nomina Anatomica Veterinaria.*

[87] This term added in accordance with the *Nomina Anatomica.*

H 26

SYSTEMA ENDOCRINA

GLANDULAE ENDOCRINAE

GLANDULA THYROIDEA

Lobus
Lobulus
Folliculus
  Endocrinocytus follicularis
  Endocrinocytus parafollicularis
  Colloidum
Glandula thyroidea accessoria
Rete capillare perifolliculare
Rete lymphocapillare perifolliculare

GLANDULAE PARATHYROIDEAE

Endocrinocytus parathyroideus
  Endocrinocytus principalis
    Endocrinocytus principalis lucidus
    Endocrinocytus principalis densus
  Endocrinocytus oxyphilicus
    [acidophilicus]

HYPOPHYSIS CEREBRI [GLANDULA PITUITARIA]

Adenohypophysis [Lobus anterior]
  Pars tuberalis
  Pars intermedia
    Endocrinocytus basophilus
    Conus adenohypophysis (su, Ru)
  Cavum hypophysis (Car, Ru, su)
  Pars distalis
    Racemus endocrinocytorum
      Endocrinocytus chromophobus
      Endocrinocytus chromophilus
        Endocrinocytus acidophilus[88]
        Endocrinocytus
          somatotrop(h)icus
        Endocrinocytus
          mammotrop(h)icus
        Endocrinocytus basophilus

Endocrinocytus
  thyrotrop(h)icus[89]
Endocrinocytus
  gonadotrop(h)icus[90]
Endocrinocytus
  corticotrop(h)icus
  Trabecula adenohypophysis
Neurohypophysis [Lobus posterior]
  Infundibulum
  Lobus nervosus
  Tractus hypothalmohypophysialis
    Gliocytus centralis [Pituitocytus]
    Substantia neurosecretoria
      Corpusculum neurosecretorium
        accumulatum
  Recessus infundibuli
Vasa sanguinea hypophysis (Homo)
  Arteria superior hypophysis
    Arteria trabecularis
  Rete capillare primarium
    Rete superficiale
      Ansa capillaris brevis
    Rete profundum
      Ansa capillaris longa
    Rete subependymale
Arteria inferior hypophysis
Vas longum portale hypophysis
Vas breve portale hypophysis
Vas capillare sinusoideum adenohypophysis
Vas capillare neurohypophysis
Vena hypophysis
Synapsis axovascularis

EPIPHYSIS CEREBRI [GLANDULA PINEALIS]

Gliocytus centralis
Endocrinocytus pinealis
  Endocrinocytus lucidus
  Endocrinocytus densus
Nervus conarius
Neurofibra perivascularis

INSULAE PANCREATICAE

Endocrinocytus pancreaticus

---

[88] These have been called "alpha cells."
[89] These have been called "beta cells."
[90] These have been called "delta cells."

Endocrinocytus alpha [Glucagonocytus]
Endocrinocytus beta [Insulinocytus]
Endocrinocytus delta

GLANDULA SUPRARENALIS [ADRENALIS]

Capsula
  Lamina fibrosa
  Lamina fibrosa
Cortex
  Zona glomerulosa[91]
  Zona fasciculata
    Pars interna
    Pars externa
  Zona reticularis
  Endocrinocytus corticalis
  Nodulus accesorius
Medulla
  Endocrinocytus medullaris
    Endocrinocytus lucidus
      [Epinephrocytus]
    Endocrinocytus densus
      [Norepinephrocytus]
  Neuronum multipolare (autonomicum)
Plexus venosus medullae
Vena centralis

PARAGANGLIA (see GLOMERA, page H 30)

Paraganglion sympathicum [-eticum]
  Glomerocytus
    Endocrinocytus granularis
    Epithelioidocytus sustentans
  Vas capillare sinusoideum
    Endotheliocytus fenestratus

THYMUS (see page H 31)

SYSTEMA CARDIOVASCULARE ET LYMPHOVASCULARE [ANGIOLOGIA]

PERICARDIUM

Pericardium fibrosum
Tela subpericardialis [subserosa]
Pericardium serosum
  Mesothelium

COR

Epicardium
  Mesothelium
Tela subepicardiaca [subserosa]
Myocardium
  Myofibra[92]
Myocytus cardiacus
Systema conducens cardiacum
  Nodus sinuatrialis
  Nodus atrioventricularis
  Myocytus nodalis
  Fasciculus atrioventricularis
    Truncus
    Crus dextrum
      Ramus cruris dextri
    Crus sinistrum
      Ramus cruris sinistri anterior
      Ramus cruris sinistri posterior
  Myofibra conducens cardiaca[93]
    Myocytus cardiacus conducens
Endocardium
  Endothelium
  Stratum subendotheliale
  Stratum myoelasticum
Tela subendocardialis
Trigonum fibrosum dextrum/sinistrum
Annulus [Anulus] fibrosus
Septum interventriculare
  Pars membranacea
  Pars muscularis

---

[91] The dog, cat and horse have a cellular arrangement in the outer cortical zone which is arcuate and hence called the *zona arcuata*. A *zona intermedia* of undifferentiated parenchyma is prominent in the dog and horse. It is located between the *zona arcuata* and *zona fasciculata*.

[92] A *Myofibra* is a series of cardiac myocytes juxtaposed end-to-end; it is not a synonym for *Myocytus*.

[93] This term in the first edition of *Nomina Histologica* also included *purkinjiensis*. The Committee dropped the latter, an adjectival construct from an eponym, in deference to the principle rejecting eponyms (*see* footnote 40).

Cartilago cordis (Car, su, eq)
Os cordis (Ru, eq)

VASA SANGUINEA

Vas capillare [Vas haemocapillare]

Endothelium
  Endotheliocytus nonfenestratus
  Endotheliocytus fenestratus
  Vesicula superficialis [Caveola]
  Membrana basalis
    Lamina lucida
    Lamina densa [basalis]
    Lamina reticularis
Pericytus [Periangiocytus]
Ansa capillaris
Vas capillare arteriale
Vas capillare venosum
Vas capillare sinusoideum[94]
  Macrophagocytus stellatus[95]
  Endotheliocytus fenestratus
  Apertura intercellularis
  Membrana basalis continua
  Membrana basalis noncontinua
Lacuna cavernosa
Rete capillare
Gemma endothelialis

Arteriae

Arteriola
  Tunica interna [intima]
    Rete elasticum
  Tunica media
  Tunica externa [adventitia]
Arteriola precapillaris [Metarteriola]
  Sphincter precapillaris
Rete arteriolare

Arteria
  Tunica interna [intima]
    Stratum subendotheliale
    Membrana elastica interna
  Tunica media
    Membrana elastica externa
  Tunica externa [adventitia]
Arteria elastotypica
  Membrana fenestrata elastica
Arteria myotypica
Arteria mixtotypica
Arteria convoluta
  Constrictor intravascularis
    Vallum musculare longitudinale
Rete arteriosum
Sinus caroticus[96]

Venae

Venula
  Tunica interna [intima]
  Tunica media
  Tunica externa [adventitia]
Venula postcapillaris
Venula colligens
Venula muscularis
Rete venosum

Vena
  Tunica interna [intima]
    Valvula
      Sinus valvulae
    Stratum subendotheliale
    Rete elasticum
  Tunica media
  Tunica externa [adventitia]
Vena myotypica
Vena fibrotypica
Rete venosum
Sinus venosus[97]

---

[94] A *Vas capillare sinusoideum* is a modified blood capillary occurring in the liver, bone marrow, and various endocrine organs.

[95] The origin and phagocytic activity of cells associated closely with the blood capillary sinusoid were concluded to be of the endothelium when resolved with light microscopy, hence the term "*Systema recticuloendotheliale.*" The *Macrophagocytus stellatus*, along with macrophages of various locations, has been, as a consequence of structure and function, included in the "macrophage system," formerly the "reticuloendothelial system" (see footnote 64).

[96] This replaces *Sinus arterialis* of the first edition of *Nomina Histologica*. With this exception, the dilated segment of the internal carotid artery, the term sinus, when applied to *Vas sanguineum*, refers to venous sinuses only.

[97] The term sinus, when applied to *Vas sanguineum*, generally refers to venous sinuses only. An exception is its preempted use in *Sinus (Bulbus) caroticus.*

Vena cavernosa
Plexus venosus

ANASTOMOSIS ARTERIOVENOSA
[ARTERIOLOVENULARIS]
Anastomosis arteriovenosa simplex
Anastomosis arteriovenosa glomeriformis
Segmentum arteriale
Pulvinar tunicae internae [intimae]
Myocytus epithelioideus
Segmentum venosum

GLOMERA (see PARAGANGLIA,
page H 28)
Glomus caroticum
Endocrinocytus granularis
Epithelioidocytus sustentans
Glomus aorticum
Glomus pulmonare

VASA LYMPHATICA
Vas lymphocapillare
Endothelium
Fibra fixationis
Rete lymphocapillare
Vas lymphaticum
Tunica interna [intima]
Endothelium
Stratum subendotheliale
Valvula
Sinus
Tunica media
Tunica externa [adventitia]
Vas lymphaticum myotypicum
Vas lymphaticum fibrotypicum
Plexus lymphaticus

VASA ET NERVI VASORUM
Vasa vasorum
Vas lymphaticum vasorum
Vas lymphocapillare vasorum
Nervi vasorum
Plexus nervorum intramuscularis
Plexus nervorum perivascularis

Plexus nervorum periarterialis
Plexus nervorum perivenosus

ORGANA HAEMOPOETICA ET
LYMPHOPOETICA

MEDULLA OSSIUM
Medulla ossium rubra
Stroma medullae
Textus haemopoeticus (see page H 11)
Textus reticularis
Vas sinusoideum
Medulla ossium flava
Medulla ossium gelatinosa
Medulla ossium fibrosa

NODUS LYMPHATICUS [LYMPHONODUS]
Capsula
Trabecula
Hilum
Vas lymphaticum afferens
Sinus lymphaticus
Sinus subcapsularis
Sinus corticalis perinodularis
Sinus medullaris
Vas lymphaticum efferens
Cortex
Nodulus lymphaticus [Lymphonodulus][98]
Nodulus primarius
Nodulus secundarius
Centrum germinale
Corona
Paracortex [Zona thymodependens]
Venula postcapillaris
Medulla
Chorda medullaris
Textus reticularis
Lymphocytus

NODUS LYMPHATICUS INVERSUS (su)

NODUS LYMPHATICUS HAEMALIS (Ru)[99]

SPLEN [LIEN]
Tunica serosa

[98] In deference to common usage "folliculus" was included in the previous *Nomina Histologica* edition
as an alternative for *Nodulus*. In this context "folliculus" is a misnomer, based upon the misconception
that nodules are cavitated.

[99] A *Nodus lymphaticus haemalis* is the "haemal node" of English literature. It is spleen-like in
organization and has lymphatic tissue, in the lymphatic sinuses of which erythrocytes normally occur.
The so-called "haemolymph" node is a lymph node that contains erythrocytes in its sinuses due to
hemorrhage in its tributary field.

Tunica fibrosa [Capsula]
Trabecula splenica
  Fibra collagena
  Fibra elastica
  Myocytus nonstriatus
Pulpa splenica
  Pulpa rubra
    Chorda splenica
  Pulpa alba
    Vagina periarterialis lymphatica
    Lymphonodulus splenicus
Arteria trabecularis
Arteria pulpae albae
  Arteria lymphonoduli[100]
Arteriosus penicillaris
  Arteriola penicillaris
  Arteriola ellipsoidea [vaginata]
  Vas capillarium terminale
Sinus venularis[101]
  Endotheliocytus fusiformis
  Fibra reticularis annularis [anularis]
Vena pulpae rubra
Vena trabecularis

THYMUS[102]
Capsula
Septum corticale
Lobulus thymi
  Cortex
  Medulla
    Epithelioreticulocytus thymi
    Corpusculum thymicum
  Thymocytus [Lymphocytus thymicus]

TONSILLA (*see* APPARATUS
  DIGESTORIA
  page H 15)

SYSTEMA NERVOSUM

MENINGES

Dura mater[103]
  Dura mater encephali
    Lamina externa
    Lamina interna
  Dura mater spinalis
Arachnoidea mater[103]
  Arachnoidea encephali
    Cisterna subarachnoidea
    Trabecula arachnoidea
    Granulatio arachnoidea
      Macula cellularis
      Colliculus cellularis
  Arachnoidea spinalis
Pia mater[103]
  Pia mater encephali
  Pia mater spinalis
    Lamina externa
      Ligamentum denticulatum
    Lamina interna
      Membrana gliae externa
Plexus choroideus
  Glomus choroideum
  Tela choroidea

PARS CENTRALIS [SYSTEMA NERVOSUM CENTRALE][104]

MEDULLA SPINALIS
Septum medianum dorsale
Fissura mediana ventralis
Sulcus lateralis dorsalis
Commissura alba
Commissura grisea
Membrana limitans gliocyti interna

---

[100] This replaces *Arteria centralis* in the first edition of *Nomina Histologica*. This artery of the white pulp is very rarely located in the center of the lymph nodule.

[101] A *Sinus venularis* is a postcapillary venule connecting the *Capillarium terminale* and *Vena pulpae rubrae*. The term *Sinus venosus* is preempted by the *Nomina Anatomica* for certain large venous channels.

[102] The *Thymus* also produces secretions called "thymosins" or "lymphopoietins" which has resulted in its classifications by some authors as an endocrine organ. A similar "secondary endocrine" function is manifest in other organs, e.g. *Ren* and *Gaster*.

[103] The *Dura mater* is also called the "pachymeninx" and the *Arachnoidea mater* and *Pia mater* are likewise, because of common origin, named the "leptomeninx," and sometimes "arachnopia."

[104] The terms under PARS CENTRALIS are new to this edition.

Canalis centralis
  Ependymocytus
Substantia alba
  Oligodendrocytus
  Neurofibra myelinata
  Funiculus dorsalis
  Funiculus lateralis
  Funiculus ventralis
Substantia grisea
  Neuronum [Neurocytus]
  Neurofibra nonmyelinata
  Neuropilus[105]
Columna grisea
  Columna dorsalis [Cornu dorsale]
  Columna lateralis [Cornu laterale]
  Columna ventralis [Cornu ventrale]
  Neuronum motorium
    Neuronum somaticum
    Neuronum autonomicum
    Neuronum fusi neuromuscularis
  Neuronum internunciale
    Neuronum commissurale
    Neuronum noncommissurale
Formatio reticularis

ENCEPHALON

CEREBELLUM
Corpus medullare
Lamina alba
Cortex cerebelli
  Stratum moleculare
    Neuronum stellatum
    Neuronum corbiferum
  Stratum neuronorum piriformium
    Corbis neurofibrarum
    Neuronum piriforme[106]
  Stratum granulosum
    Neuronum stellatum magnum[107]
    Neuronum granuliforme
    Glomerulus
  Neurofibra muscoidea
  Neurofibra ascendens
  Neurofibra parallela

CEREBRUM

CORTEX CEREBRI
Stratum moleculare [plexiforme]
  Neuronum horizontale
Stratum granulare externum
  Neuronum pyramidale parvum
Stratum neuronorum pyramidalium
  externum
  Neuronum pyramidale magnum
Stratum neuronorum pyramidalium
  externum
  Neuronum pyramidale magnum
  internum
  Neuronum pyramidale medium
  Neuronum pyramidale magnum
Stratum neuronorum fusiformium
  Neuronum fusiforme parvum
  Neuronum fusiforme medium

PARS PERIPHERICA [SYSTEMA
NERVOSUM PEIPHERICUM][108]

Ganglion spinale (sensorium)
  Neuronum pseudounipolare
  Gliocytus ganglii[109]
  Capsula
Ganglion autonomicum [viscerale]
  Neuronum multipolare
    Glomerulus dendriticus
    Synapsis axodendritica
    Synapsis axosomatica
  Gliocytus ganglii[109]
  Capsula
Nervus
  Neurofibra
    Neurofibra myelinata
    Neurofibra nonmyelinata
    Neurofibra afferens
    Neurofibra efferens
    Neurofibra autonomica
      Neurofibra preganglionica
      Neurofibra postganglionica
    Plexus nervorum autonomicus

---

[105] The *Neuropilus* is the meshwork formed of axons, dendrites, and gliocyte processes.
[106] These were formerly called "Purkinje cells."
[107] These were formerly called "Golgi cells."
[108] For visceral efferent components of the PARS PERIPHERICA refer to PARS AUTONOMICA of the *Nomina Anatomic* (page A 11). This essentially functional grouping of efferent neurons within both the PARS CENTRALIS and PARS PERIPHERICA does not receive a separate listing in the *Nomina Histologica*.
[109] These have also been called "neurosatellite cells" or "satellite cells" (*see* footnote 46).

Endoneurium
Perineurium
  Pars fibrosa
  Pars epithelioidalis
Epineurium
  Epineurium superficiale
  Epineurium profundum

ORGANA SENSUUM [SENSORIA]

Epithelium sensorium
  Epitheliocytus neurosensorius[110]
  Epitheliocytus sensorius[111]
  Epitheliocytus sustentans

ORGANUM VISUS [VISUALE]

OCULUS
Nervus opticus

BULBUS OCULI

Tunica fibrosa bulbi
  Sclera
    Lamina episcleralis
    Substantia propria
    Lamina fusca
      Melanocytus
    Area cribrosa
    Annulus (anulus) sclerae
  Cornea
    Epithelium anterius
    Lamina limitans anterior
    Substantia propria
    Lamina limitans posterior
    Epithelium posterius
    Limbus
    Angulus iridocornealis
      Ligamentum pectinatum
      Spatia anguli iridocornealis
      Reticulum trabeculare

Tunica vasculosa bulbi [Uvea]
  Choroidea
    Melanocytus choroideus
    Lamina suprachoroidea

Spatium perichoroideale
Substantia propria
Tapetum lucidum (Car, Ru, eq)
  Tapetum fibrosum (Ru, eq)
  Tapetum cellulosum (Car)
Lamina choroidocapillaris
Complexus basalis
  Stratum elasticum
  Stratum fibrosum
  Lamina basalis
Corpus ciliare
  Stratum musculare
    Musculus ciliaris
      Fibrae meridionales
        [longitudinalis]
      Fibrae radiales
      Fibrae circulares
  Stratum vasculosum
  Pars ciliaris retinae
    Epithelium pigmentosum
    Epithelium nonpigmentosum
Iris
  Granulum iridicum (Ru, eq)
  Stroma
    Stratum nonvasculosum
    Stratum vasculosum
  Musculus sphincter pupillae
  Pars iridica retinae
    Myopigmentocytus iridicus
    Musculus dilator pupillae
    Epithelium posterius
      pigmentosum
    Pigmentocytus

Tunica interna [sensoria] bulbi
  Retina
    Pars iridica retinae
    Pars ciliaris retinae
    Ora serrata
    Pars optica retinae
      Stratum pigmentosum
        Pigmentocytus
      Stratum nervosum
        Stratum neuroepitheliale
          [photosensorium]
        Epitheliocytus (neurosensorius)
          bacillifer
          Segmentum internum
            Cilium

[110] *Epitheliocytus neurosensorius* is a nerve cell modified as a receptor for sight or olfaction.
[111] *Epitheliocytus sensorius* is a non-neuronal cell modified for reception of specific stimuli as in the organs for hearing, equilibration, and taste.

H 33

Segmentum externum
Discus membranaceus
Epitheliocytus (neurosensorius)
conifer
Segmentum internum
Cilium
Segmentum externum
Discus membranaceus
Stratum limitans externum
Stratum nucleare externum
Corpus epitheliocyti
neurosensorii
Processes preterminalis
Stratum plexiforme externum
Spherula terminalis epitheliocyti
bacilliferi
Pes terminalis epitheliocyti
coniferi
Stratum nucleare internum
Neuronum horizontale
Neuronum bipolare
Processus preterminalis
Neuronum amacrinum
Gliocytus radialis
Processus radialis
Stratum plexiforme internum
Stratum ganglionare
Neuronum multipolare
Stratum neurofibrarum
Astrocytus protoplasmaticus
Stratum limitans internum
Camera anterior
Camera posterior
Camera vitrea
Corpus vitreum
Canalis hyaloideus
Lens
Fibra centralis
Fibra transitoria
Fibra principalis
Epithelium lentis
Capsula lentis
Radius lentis
Sutura lentis
Zonula ciliaris
Fibra zonularis
Spatium zonulare

ORGANA OCULI ACESSORIA

Palpebra
Cilium

Glandula tarsalis
Glandula ciliaris
Glandula sebacea

Tunica conjunctiva
Epithelium
Glandula conjunctivalis
Tela subconjunctivalis
Palpebra tertia [Membrana nictitans] (Car,
Un)
Cartilago
Glandula superficialis
Glandula profunda
Noduli lymphatici aggregati

Apparatus lacrimalis
Glandula lacrimalis
Stroma
Parenchyma
Lacrimocytus
Myoepitheliocytus

ORGANUM VESTIBULOCOCHLEARE

AURIS INTERNA

LABYRINTHUS MEMBRANACEUS

Labyrinthus vestibularis
Spatium endolymphaticum
Cellula sensoria pilosa
Epitheliocytus piriformis
Epitheliocytus columnaris
Epitheliocytus sustentans
Macula sacculi
Macula utriculi
Membrana statoconiorum
Statoconium
Ampulla membranacea
Crista ampullaris
Cupula gelatinosa
Planum semilunatum
Ductus semicircularis
Ductus endolymphaticus
Saccus endolymphaticus
Ductus perilymphaticus
Trabecula perilymphatica

Labyrinthus cochlearis
Spatium endolymphaticum
Ductus cochlearis

Paries internus
  Foramen nervosum
  Limbus laminae spiralis
    Labium limbi tympanicum
    Sulcus spiralis internus
    Labium limbi vestibulare
      Dentes acustici
      Epitheliocytus interdentalis
    Membrana tectoria [gelatinosa]
Paries externus
  Crista spiralis [Ligamentum
    spirale][112]
  Crista basilaris
  Sulcus spiralis externus
  Prominentia spiralis
  Vas prominens
  Stria vascularis
Paries vestibularis [Membrana
    vestibularis]
Paries tympanicus [Membrana spiralis]
  Lamina basilaris
  Zona arcuata
  Zona pectinata
  Vas spirale
Organum spirale
  Epitheliocytus sustentans internus
  Epitheliocytus limitans internus
  Epitheliocytus sensorius pilosus
      internus [piriformis]
  Fasciculus spiralis internus
      neurofibrarum
  Epitheliocytus phalangeus internus
  Epitheliocytus pilaris internus
  Cuniculus internus
    Neurofibra radialis
    Neurofibra spiralis
  Epitheliocytus pilaris externus
  Cuniculus medius
  Epitheliocytus sensorius pilosus
      externus [columnaris]
  Fasciculus spiralis externus
      neurofibrarum
  Epitheliocytus phalangeus externus
    Processus phalangeus
  Cuniculus externus
  Epitheliocytus limitans externus
  Epitheliocytus sustentans externus
  Membrana reticularis
  Membrana tectoria [gelatinosa]

Spatium perilymphaticum
  Scala vestibuli
  Scala tympani
  Helicotrema

Labyrinthus osseus
Vestibulum
  Fenestra vestibuli [Fenestra ovalis]
  Canalis semicircularis osseus
Cochlea
  Fenestra cochleae [Fenestra rotunda]
  Modiolus
    Canalis spiralis
      Ganglion spirale
      Fasciculus spiralis neurofibrarum
    Canalis longitudinalis
      Pars cochlearis nervi
        vestibulocochlearis
    Lamina spiralis

AURIS MEDIA

Cavitas tympanica
Tunica mucosa
Membrana tympani secundaria
Membrana tympani
  Stratum mucosum
  Stratum cutaneum
  Annulus [Anulus] fibrocartilagineus
  Pars flaccida
  Pars tensa
    Stratum circulare
    Stratum radiatum
Ossicula auditus

Tuba auditiva [auditoria]
Tunica mucosa
  Epitheliocytus ciliatus
  Exocrinocytus caliciformis
  Glandula tubaria
Cartilago tubae auditivae
Noduli lymphatici aggregati
Diverticulum tubae auditivae [*auditoriae*]
  (eq)

AURIS EXTERNA
Glandula ceruminosa
Glandula sebacea

---

[112] This term is new to this edition and conforms with the *Nomina Anatomica*, page A 81.

ORGANUM OLFACTUS

Tunica mucosa olfactoria
  Epithelium olfactorium
    Epitheliocytus neurosensorius
      olfactorius
      Dendritum
      Bulbus dendriticus
      Cilium
      Axon [Neurofibra olfactoria]
    Epitheliocytus sustentans
    Epitheliocytus basalis
  Glandula olfactoria
Organum vomeronasale
  Ductus vomeronasalis
  Cartilago vomeronasalis
  Tunica mucosa glandularis
      [vomeronasalis]
  Tunica mucosa olfactoria

ORGANUM GUSTUS
[GUSTATORIUM]

Caliculus gustatorius [Gemma gustatoria]
  Porus gustatorius
  Epitheliocytus sensorius gustatorius
    Microvillus
  Epitheliocytus sustentans
  Epitheliocytus basalis
Plexus subcalicularis neurofibrarum
Neurofibra gustatoria

INTEGUMENTUM COMMUNE

CUTIS

Epidermis
Stratum corneum
  Stratum disjunctum
  Squama cornea
  Epitheliocytus squamosus
Stratum lucidum
  Epitheliocytus squamosus
Stratum granulosum

  Epitheliocytus squamosus
  Granulum keratohyalini
  Granulum lamellosum
Stratum spinosum
  Epitheliocytus spinosus
    Tonofibrilla
    Tonofilamentum
    Macula adherens [Desmosoma]
Stratum basale
  Epitheliocytus basalis
  Melanocytus [Melanoblastocytus][113]
    Premelanosoma
    Melanosoma [Granulum melanini]
  Melanophorocytus[113]
  Macrophagocytus intraepidermalis[114]
  Epithelioidocytus tactus[115]

Dermis [Corium]
Stratum papillare
  Papilla
Stratum reticulare
Ansa capillaris intrapapillaris
Rete arteriosum subpapillare
Rete arteriosum dermidis
Plexus venosus subpapillaris superficialis
Plexus venosus subpapillaris profundus
Plexus venosus dermidis profundus
Rete lymphocapillare cutis profundum
Plexus nervorum subepidermidis
Plexus nervorum dermatis
Terminatio nervi cutis
    (*see* TERMINATIONES
    NERVORUM, page H 13)

Tela subcutanea
Panniculus adiposus
Plexus venosus subcutaneus
Plexus lymphaticus subcutaneus
Rete lymphocapillare subcutaneum
Plexus nervorum subcutaneus
    (*see* TERMINATIONES
    NERVORUM, page H 13)

Pilus
Apex (pili)
Scapus (pili)

---

[113] The *Melanocytus* is also called a *Melanoblastocytus*, the cell producing melanin. Some would reserve the latter term for the less differentiated, dividing, precursor cell of neural crest or neural tube origin. Phagocytes of epithelial and connecting tissues which ingest melanin are *Melanophorocyti*.

[114] This term replaces *Dendrocytus granularis nonpigmentosus* of the first edition. This cell was formerly called a "cell of Langerhans" and was believed to be an effete melanocyte.

[115] This cell was formerly called a "cell of Merkel."

Radix (pili)
Bulbus (pili)
  Cervix
  Cavitas
Medulla (pili)
  Epitheliocytus polyhedralis
    Granulum trichohyalini
    Tonofibrilla
    Tonofilamentum
    Granulum melanini
Cortex (pili)
  Granulum melanini
Cuticula (pili)
  Epitheliocytus cuticularis
Folliculus (pili)
  Cervix
  Fundus
  Canalis
Vagina epithelialis radicularis
  Vagina epithelialis radicularis interna
    Cuticula vaginalis
      Epitheliocytus cuticularis
    Stratum epitheliale internum
      [granuliferum]
    Stratum epitheliale externum
      [pallidum]
  Vagina epithelialis radicularis externa
  Membrana basalis [vitrea]
Vagina dermalis radicularis
  Stratum circulare internum
  Stratum longitudinale externum
Papilla (pili)
  Collum papillae
Matrix (pili)
  Epitheliocytus matricis
  Melanophorocytus
  Melanocytus
Musculus arrector pili
Pilus tactilis (Car, Un)[116]
  Sinus sanguineus folliculi
Folliculus complexus (Car, ov)[117]
  Folliculus pili primarius
  Folliculus pili secundarius
Pilus claviformis
Rete capillare papillae pili
Rete capillare vaginae dermalis radicularis
Plexus nervorum circularis

UNGUIS (HOMO)
Corpus
  Lunula
  Facies externa
  Facies interna
Stratum corneum
Lectulus
  Crista
  Sulcus
Stratum germinativum (unguis)
  Stratum spinosum
  Stratum basale
Dermis
  Crista dermalis (unguis)
  Sulcus dermalis (unguis)
Matrix
  Crista matricis
  Sulcus matricis
  Sinus
Hyponychium
Eponychium
Vallum
  Vallum laterale
  Vallum posterius

UNGUICULA (CAR) / UNGULA (UN)
Epidermis
  Epidermis tubularis (Un)
    Tubulus epidermalis
      Epidermis suprapapillaris
      Epidermis peripapillaris
    Epidermis intertubularis
  Epidermis lamellata
    Lamella epidermalis primaria [cornea]
    Lamella epidermalis secundaria (eq)
  Epidermis limbi [Perioplum (Un)]
Dermis [Corium]
  Dermis papillaris
    Papilla dermalis
  Dermis lamellata
    Lamella dermalis primaria
    Lamella dermalis secundaria (eq)

CORNU (RU)
Epidermis
  Tubulus epidermalis
  Epidermis intertubularis

[116] This is a prominent hair specialized for sensory reception.
[117] A follicle complex consists of a single primary follicle and several secondary follicles. The latter branch from the primary follicle, so that all hairs emerge from a common opening.

Epiceras
Dermis [Corium]
  Papilla dermalis

GLANDULAE CUTIS
Glandula sudorifera
  Glandula sudorifera apocrina
    Portio terminalis
      Alveolus
        Exocrinocytus
        Myoepitheliocytus fusiformis
    Ductus sudorifer
  Glandula sudorifera merocrina [eccrina]
    Portio terminalis
      Alveolus
      Acinus
        Exocrinocytus lucidus
        Exocrinocytus densus
        Myoepitheliocytus fusiformis
      Ductus glandularis
      Porus glandularis
Glandula sebacea
  Glandula sebacea pili
  Glandula sebacea libera
  Sacculus sudorifer
    Exocrinocytus sebaceus [Sebocytus]
  Ductus glandularis

MAMMA

Stratum adiposum

Stratum fibrosum
Glandula mammaria
  Lobus
  Lobulus
    Septum interlobulare
    Alveolus glandulae
      Exocrinocytus lactus [Lactocytus]
        Gutta adipis
        Granulum proteini
      Myoepitheliocytus stellatus
    Ductus alveolaris lactifer
  Ductus lactifer
  Ductus lactifer colligens
  Sinus lactifer
    Pars glandularis (Ru, eq)
    Plica annularis [anularis] mucosae
      (Ru)
    Pars papillaris (Ru, eq)
  Papilla mammae
    Ductus papillaris (Ru, eq)
      Musculus sphincter papillae
    Ostium papillare
Areola
  Glandula areolaris
Rete lymphaticum intralobulare
Rete lymphaticum interlobulare
Plexus nervorum interlobularis
Plexus nervorum parapapillaris

NOMINA EMBRYOLOGICA

Prepared by
a Subcommittee of the
International Anatomical Nomenclature Committee
appointed at the Seventh International Congress
of Anatomists in New York, 1960,
and based upon a Provisional Version approved by the
Ninth International Congress in Leningrad, 1970,
being further revised with the participation of the
International Committee on Veterinary Anatomical Nomenclature
and approved by the
Tenth International Congress of Anatomists
at Tokyo, 1975
and being further revised and approved by the
Eleventh International Congress of Anatomists
at Mexico City, 1980

CONTENTS

CONTENTS

CONTENTS

E vi

STYLE USAGE

1. Terms within squared brackets are officially recognized alternatives or synonyms. Examples:

   > Pregnantia [Graviditas]
   > Ventriculus [Gaster] primitivus
   > Neuroporus rostralis [cranialis]

   The second and third examples illustrate alternatives for a part only of complete terms.

2. Terms for some structures which are bilateral, and must therefore be qualified as right (dexter) or left (sinister), are shown as follows:

   > Atrium dextrum/sinistrum

3. Terms are placed in rounded brackets for three purposes.

   Firstly, to indicate unofficial but familiar alternatives. Examples:

   > Reproductio sexualis (gametica)
   > Mesoderma splanchnicum (viscerale)

   Secondly, to indicate additional components of some terms which are frequently omitted. Example:

   > Gametocytus primarius (I)

   Thirdly, to enclose the terms *partim* (= in part, partly) and *plerumque* (= largely, mostly) which indicate that a structure is partly or preponderantly derived from a precursor. Examples:

   > Ansa umbilicalis intestini
   >   Colon transversum (partim)
   > Viscerocranium
   >   Maxilla (plerumque)

4. Terms are put in the singular number when parts are single and nonsegmental, or when bilaterally paired. Examples:

   > Notochorda
   > Placoda nasalis

5. Terms are put in the plural when parts are in serial order, or when occurring in irregular local groups. Examples:

   > Somiti
   > Tubuli seminiferi

6. Compound words in which adjoining letters resemble a diphthong or repeat a letter are separated by a hyphen. Examples:

   > Meso-esophageum
   > extra-embryonicus
   > Pre-enteron

INTRODUCTION TO SECOND EDITION

Subsequent to the Seventh International Congress of Anatomists, held at New York in 1960, the International Anatomical Nomenclature Committee (I.A.N.C.) authorized its Honorary Secretary, Professor G. A. G. Mitchell, to establish a Subcommittee on Embryology for the purpose of preparing a set of embryological terms suitable to become the basis of a standard, international terminology. To this end Professor Leslie B. Arey was asked to serve as Convener of the Subcommittee and he, in turn, invited Professor Harlan W. Mossman to serve as its Honorary Secretary. Representatives of various countries and languages were then recruited as other members of the Subcommittee. (Original representatives were: Professors Rodolfo M. Amprino, Italy; B. J. Anson, U.S.A.; R. J. Blandau, U.S.A.; D. H. Bucklin, U.S.A.; Liza W. Chacko, India; J. G. Forsberg, Norway; W. J. Hamilton, G.B.; M. Niizima, Japan; F. Orts-Llorca, Spain; C. P. Raven, The Netherlands; P. Sengel, France; F. Strauss, Switzerland. These were later joined by Professors F. D. Allan, U.S.A.; B. I. Balinsky, S. Africa; A. M. Dalcq, Belgium; T. W. Glenister, G.B.; C. M. Goss, U.S.A.; A. G. Knorre, U.S.S.R.; Z. Mahran, Egypt; and A. F. Weber, U.S.A.)

Initially, the Convener and Honorary Secretary composed tentative lists of terms pertaining to the several fields dealing with development, which were revised and repeatedly extended. In 1967 the assembled report was distributed to members of the Subcommittee, and sent to about 175 other embryologists. A request for criticisms and suggestions elicited about forty responses. In June, 1968, members of the Subcommittee met in London to engage in further review and revision, the Ciba Foundation of London generously acting as host. This meeting was of great importance, since it produced a drastically altered and significantly improved compilation. An omission, due to lack of time, was satisfactory organization of the original alphabetical list of experimental terms into more useful categories. Hence the approval of the report on *Nomina Embryologica* by the Ninth International Congress of Anatomists, meeting in 1970 at Leningrad, included the understanding that such a revised list would be presented for approval to the Tenth Congress in 1975. (The provisional list presented at Leningrad by Professors Arey and Mossman was arranged in parallel columns—in Latin and English vernacular. Professor Howe, University of Wisconsin, advised on problems of Latinization. The National Library of Medicine of America kindly financed the copies of this list distributed at Leningrad.) The London meeting was attended by Professors Arey, Balinsky, Chacko, Dalcq, Glenister, Goss, Mossman, Strauss and Weber.

In August, 1974, representatives of the Subcommittee met as guests of the Kroc Foundation, Santa Ynez, California, to review and revise further the current edition of *Nomina Embryologica*. Those responding to invitations to attend these sessions were: F. D. Allan, U.S.A.; L. B. Arey, U.S.A.; J. Faber, The Netherlands; D. Rudnick, U.S.A.; F. Strauss, Switzerland; and R. L. Zwemer, U.S.A. Invited specialists were Professor Robert L. Bacon of the University of Oregon, U.S.A., and Professor Caroline M. Czarnecki of the University of Minnesota, U.S.A., whose contributions are acknowledged with gratitude. Aided by numerous suggestions received since the publication of the original report, this working group made many revisions and also simplified numerous Latin terms into more easily understood synonyms. This revision was approved by the Tenth International Congress, meeting in 1975 at Tokyo. Continued labors toward completing a satisfactory organization and extension of the lists of experimental terms failed to produce a definitive report ready for submission to the Tenth

Congress. (This list has also proved difficult to express in Latin, but it is hoped that it will eventually appear in English as a glossary.)

A meeting of the Subcommittee, for further revising, was scheduled in conjunction with the International Morphological Congress at Toledo, U.S.A. in the summer of 1979, but it failed to produce a working group. The numerous changes incorporated into the current edition came about through memoranda emanating from the 1979 meeting of the I.A.N.C. in London, from a complete review by the Convenor, Professor L. B. Arey, and from suggestions by Professors J. McKenzie, F. Strauss, T. Mori, T. Donath, R. O'Rahilly, T. Tobias, R. Hullinger, and H. Elias. The final revision was approved in 1980 by the Eleventh International Congress of Anatomy at Mexico City.

It is obvious that the preparation of a nomenclature for embryology presents organizational difficulties not inherent in gross anatomy or histology. This is because, to be useful, embryological lists must deal with developmental processes and time sequences, along with factors of position and the progressive subdivision of parts. Incidentally, because of such groupings some terms have to appear more than once; for example, "somite" must occur under the history of the mesoderm, the skeletal system and the muscular system. An alternative to an organization by categories would be a simple alphabetical list, but such listing would offer no clues to the relations existing between different terms or to where in the list to find an unknown or unrecalled term.

In general, the style and arrangement used in *Nomina Anatomica* have been followed. Departures occur because many topics not pertinent to gross anatomy are integral parts of *Nomina Embryologica*. For the most part items are placed in alphabetical order, but in some instances a more logical sequence seemed preferable. It was agreed the *Nomina Embryologica* should include general developmental terms and concepts, but that in specifics it should be restricted to amniotes, with primary emphasis on mammals and man. Names of structures derived from more than the obviously designated precursor are followed by *partim* (= partly) or *plerumque* (= mostly, for the most part).

The Subcommittee is fully cognizant of the fact that this third revised list is only a renewed attempt toward producing a standard terminology in a discipline whose general organization and detailed handling are subject to variant thought. Indubitably it will, in the future, undergo further refinements, just as the original Basle *Nomina Anatomica*, of 1895, proved to be a provisional document.

Professor LESLIE B. AREY
Convener,

Professor DOROTHEA RUDNICK
Honorary Secretary,
Subcommittee for Embryological Terms,
International Anatomical Nomenclature Committee

The names and addresses of present members of the Subcommittee are given below, but correspondence containing questions, criticisms or suggestions should be addressed preferably to the Subcommittee's Convener, Professor Arey, its Secretary, Dr. Rudnick, or to the Honorary Secretary of the I.A.N.C.

The secretaries of all known anatomical societies, associations, and federations have also been asked by the Honorary Secretary, I.A.N.C., to act as intermediaries in forwarding to him comments on any of the nominal lists, including *Nomina Embryologica*, which may be sent to them by colleagues in their own countries. Use of this channel of communication may obviate linguistic difficulties. Some countries have already instituted committees, associated with their own national societies, for the specific purpose of continuous assessment of nomenclature. The I.A.N.C. hopes that more of such national groups will appear; they provide not only an excellent channel for transmission or revisional comments, but also a local forum for preliminary discussion and modification of the suggestions of individual scientists. Thus, there exists a number of channels along which any individual can express opinions. Whichever is used, delay is to be avoided.

The next opportunity for any changes in *Nomina Embryologica* will be at the Twelfth International Congress of Anatomists in England during the summer of 1985, and all suggestions for such changes must be collected, edited, and circulated to all I.A.N.C. members (and sometimes to others for expert opinions) in order to prepare an agenda for consideration at the World Congress and presentation at a plenary session for final ratification. Since these processes occupy several months, all amendments should be sent in at least five months before the date of the London (1985) Congress.

ROGER WARWICK

Honorary Secretary,
International Anatomical Nomenclature Committee,
℅ The Department of Anatomy, Guy's Hospital Medical School,
London SE1 9RT, Great Britain

INTERNATIONAL ANATOMICAL NOMENCLATURE COMMITTEE
EMBRYOLOGY SUBCOMMITTEE

(*see pages A2–A8 for complete addresses.*)

PROF. F. D. ALLAN
PROF. R. M. AMPRINO
PROF. L. B. AREY, (*Convener*)
PROF. B. I. BALINSKY
PROF. N. BJORKMAN
DR. J. FABER
PROF. J. FAUTREZ
PROF. T. W. GLENISTER
PROF. H.-J. KRETSCHMANN
DR. J. MCKENZIE
PROF. Z. MAHRAN
*PROF. E. MEITNER
PROF. V. MONESI
PROF. H. NISHIMURA
PROF. R. O'RAHILLY
DR. D. RUDNICK, *Secretary*
*PROF. K. THEILER
PROF. A. F. WEBER

* Appointed since Mexico revision.

REPRODUCTIO

NOMINA GENERALIA

Modi reproductionis
  Reproductio sexualis (gametica)
  Reproductio asexualis (agametica)
    Gemmatio
    Fissio
    Fragmentatio
    Polyembryonia
    Strobilatio
  Oviparitas
  Ovoviviparitas
  Viviparitas
  Autogamia
  Isogamia
  Heterogamia
  Androgenesis
  Gynogenesis
  Parthenogenesis
    naturalis
    artificialis
    facultativa
Cursus reproductionis
  Ovulatio
    uniovulatoria
    multiovulatoria (multiplex)
    spontanea
    superovulatoria [superovulatio]
    inducta
  Ovipositio
  Copulatio
    Coitus
    Ejaculatio
  Inseminatio
    Semen
  Amplexus
  Fertilisatio
    Fecundatio
    Superfecundatio
    Superfetatio
  Fissio
  Gastrulatio
  Neurulatio
  Embryogenesis
  Paedogenesis [Pedo-]

Neotaenia [-tenia]
Cycli genitales feminini
  Cyclus ovaricus
    Phasis ovogeneticus
    Phasis follicularis
    Phasis involutionis
  Typus monoestrosus
  Typus polyoestrosus
  Cyclus oestrosus
    Pro-oestrus[1]
    Oestrus [Estrus][2]
    Metoestrus
    Dioestrus
    Anoestrus
  Cyclus menstrualis
    Typus ovulatorius
      Phasis follicularis
      Phasis lutealis (progesteronis)
      Phasis ischaemica [ischemica]
      Phasis menstrualis
        Menses [Catamenia]
      Phasis postmenstrualis
    Typus anovulatorius
  Amenorrhea
  Menarche
  Climacter
    Menopausa
Pregnantia [Graviditas]
Cyclus pregnantiae
  Conceptio
  Conceptus[3]
  Abortus
  Periodus tubalis
  Periodus uterina
    Phasis preimplantationalis
    Phasis implantationalis
    Phasis placentalis
    Terminus
  Parturitio
    Partus
    Puerperium
Pseudopregnantia
Cyclus mammarius
  Phasis inactiva
  Phasis proliferativa
  Lactatio
    Phasis colostralis
    Phasis lactifera

---

[1] *Pro-* has been adopted as a prefix to signify "before" in time. *See pre-*, page E 9.

[2] *Oestrus* and all its combinations may also be spelled *estrus*. The "e" is long, which is emphasized by retaining the diphthong.

[3] *Conceptus* is an inclusive term denoting the embryo and all its membranes.

Phasis involutionis
Cyclus vaginalis
  Phasis incornificiens
  Phasis cornificiens
  Phasis desquamationis
Pregnantia [Graviditas]
  Pregnantia uterina
  Situ cornuali [Situs cornualis]
    Situ fundali [Situs fundalis]
    Situ corporali [Situs corporalis]
  Pregnantia extrauterina [ectopica]
    primaria
    secundaria
    ovarica
    abdominalis
    tubalis
    interstitialis
  Gestatio
    monembryonica
    diembryonica
    polyembryonica
  Superfetatio
  Periodus gestationis
  Paritas
    Primiparitas
    Multiparitas
    Nulliparitas
Cycli genitales masculini
  Periodus libidinis
  Phases testiculares
    infantilis
    prepubertalis
    pubertalis
    matura
  Cyclus spermatogenicus
  Involutio

GAMETOGENESIS

Spermatogenesis[4]
  Cellulae sustentaculares
  Unda spermatogenica
  Cyclus spermatogenicus
    Spermatogonium
      Spermatogonium A
      Spermatogonium B
     Corpus chromatoideum
     Status plasmodiatis

Spermatocytus primarius
  Idiosoma
Spermatocytus secundarius
Spermatidium
Spermiogenesis
  Corpus chromatoideum
  Complexus golgiensis
  Acrosoma [Vesicula acrosomalis]
  Granulum proacrosomale
  Granulum acrosomale
  Flagellum
  Spermatozoön [Spermium][5]
Ovogenesis[4]
  Cyclus ovogeneticus
    Ovogonium
    Ovocytus primarius (I)
      Polus animalis
      Polus vegetalis
    Corpus polare primum (I)
    Ovocytus secundarius (II)
    Corpus polare secundum (II)
    Ovum [Ovum maturum][5]

MEIOSIS

Phenomena nuclearia
  Ploideae
    Euploidea
    Aneuploidea
    Heteroploidea
    Diploidea
    Triploidea
    Tetraploidea
    Polyploidea
  Divisiones maturationis
    Divisio equalis
    Divisio reductans
      Divisio prereductans
      Divisio postreductans
    Gametocytus primarius (I)
      Phasis intermedia
      Phasis proleptotenica
      Phasis leptotenica
      Phasis synapsalis
      Phasis zygotenica
      Phasis pachytenica
      Phasis diplotenica
        Chiasmata

---

[4] For details of maturational stages, *see* MEIOSIS, page E 6.
[5] For details, *see* GAMETI, page E 7.

Diakinesis
  Bivalentia
  Phasis mitotica
Gametocytus secundarius (II)
  Phasis mitotica
Gametus
  Chromosoma univalens
    Actosoma
    Gonosoma
      X-chromosoma
      Y-chromosoma

GAMETI [GONOCYTI]

Spermatozoön [Spermium]
  Caput
    Nucleus
    Acrosoma
    Cytoplasma pericephalicum
      Perforatorium
  Cervix
    Lamina articularis
    Centriolum proximale
Cauda [Flagellum]
  Cytoplasma periflagellatum
  Centriolum distale
  Pars media
    Axonema [Filamentum axiale]
    Vagina mitochondrialis
    Annulus [Anulus]
  Pars principalis
    Axonema [Filamentum axiale]
    Fibrillae densae externae
    Vagina fibrosa
  Pars terminalis
    Axonema [Filamentum axiale]
Ovum (Ovotidium)
  Involucra
    Chorion
    Membrana vitellina
    Tunica mucina
    Spatium perivitellinum
    Zona pellucida [Membrana pellucida]
  Ovoplasma
    Ovolemma
    Cortex
      Granula corticalia
    Deuteroplasma [Vitellus]

Ovonucleus
Formae ovi
  Ovum oligolecithale
  Ovum mesolecithale
  Ovum megalecithale
  Ovum isolecithale
  Ovum telolecithale
  Ovum centrolecithale
  Ovum cleidoicum

ONTOGENESIS

Gameti
  Ovum
  Spermatozoön [Spermium]
Maturatio gametorum[6]
Ovulatio
Ejaculactio
Fertilisatio
  Penetratio spermi
  Conus fertilisationis
  Filamentum acrosomale
  Perforatorium
  Via spermatica
  Via penetrativa
  Via copulativa
  Membrana vitellina
  Membrana fertilisationis
  Spatium perivitellinum
  Liquor perivitellinus
  Micropylum
  Monospermia
  Dispermia, trispermia, et cetera
  Polyspermia
  Aster spermaticus
  Pronucleus masculinus
  Pronucleus femininus
  Zygota
Fissio
  asymmetrica
  bilateralis
  radialis
  spiralis
    dextra
    sinistra
  Fissio totalis
    equalis

---

[6] For details, *see* MEIOSIS, page E 6.

inequalis
Fissio partialis
  discoidalis
  superficialis
Fissio determinata
Fissio indeterminata
Planum fissionis
  equatoriale
  latitudinale
  meridionale
Nucleus fissionis
Via fissionis
Blastomerus
  Macromerus
  Mesomerus
  Micromerus
Morula
Blastula
  Blastocoelia[7]
  Stereoblastula
  Discoblastula
    Zona marginalis
  Coeloblastula
  Blastocystis[8]
Gastrulatio
  Stratificatio germinalis
  Motus morphogenetici
    Ingressio
    Immigratio
    Invaginatio
    Involutio
    Epibolia
    Embolia
    Convergentia
    Elongatio
    Delaminatio
Gastrula
  Blastoporus
    Labia blastoporalis
      Pars dorsalis
      Pars lateralis
      Pars ventralis
    Embolus vitellinus
  Archenteron
    Strata germinalia
  Cavitas subgerminalis
  Epiblastus
  Hypoblastus

Mesoblastus
  Chordamesoderma
  Mesenchyma
  Mesoderma
Ectoderma
  Neurectoderma
Endoderma
  Lamina prechordalis
  Mesendoderma
Discus embryonicus
  Area opaca
  Area pellucida
  Area formativa
Neurulatio
Neurula
  Lamina neuralis
  Plicae neurales
  Strata germinalia
Periodus embryonica
  Periodus ante discum embryonicum
  Periodus post discum embryonicum

Periodus fetalis
Periodus perinatalis
  Periodus pronatalis
  Periodus postnatalis
Etas fetalis [Aetas]
  Etas menstrualis
  Etas ovulationis
  Etas coitus
  Etas fecundationis
Ovum
Embryo
  Primordium
  Gemma
Fetus
Neonatus
Metamerismus
Branchiomerismus
Plasma somaticum (Somatoplasma)
Plasma germinale
Embryogenesis
  Morphogenesis
  Organogenesis
  Inductio
  Differentiatio
    Cytogenesis
    Histogenesis

---

[7] The root is from the Greek *koilia* meaning cavity. In the following lists it is spelled "-coel-" to distinguish it from the root form "-cele", from the Greek *kele* meaning hernia or a tumor.
[8] For details, *see* MORPHOGENESIS, page E 9.

MORPHOGENESIS

Pedunculus connexens [P. corporis]

PERIODUS ANTE DISCUM EMBRYONICUM

Fissio
Morula
Blastocystis unilaminaris
  Trophoblastus
  Blastocoelia
  Massa cellularis interior (Embryoblastus)
Blastocystis bilaminaris
  Trophoblastus
  Cavitas amniotica
  Endoblastus extra-embryonicus
  Discus embryonicus
  Saccus vitellinus primaria
Blastocystis trilaminaris[9]
  Trophoblastus
    Cytotrophoblastus
    Syncytiotrophoblastus
  Endoblastus extra-embryonicus
  Mesoblastus extra-embryonicus
  Epiblastus embryonicus
  Endoderma embryonicum
  Saccus vitellinus definitivus
  Amnion primarium

PERIODUS DISCI EMBRYONICI

Membrana oropharyngealis
    [buccopharyngealis]
  Lamina prechordalis[10]
Linea primitiva
  Sulcus primitivus
  Nodus primitivus
  Fovea primitiva
  Processus notochordale
    Lamina notochordalis
    Canalis notochordalis
Ectoderma embryonicum
Mesoderma intra-embryonicum
Endoderma embryonicum
Tectum sacci vitellini

PERIODUS INITIALIS SULCI NEURALIS

Canalis neurentericus
Lamina neuralis
  Sulcus neuralis
Junctio neuro-ectodermalis
Mesoderma paraxiale [Epimerus]
Mesoderma intermedium [Mesomerus]
Mesoderma laminae lateralis
    [Hypomerus]
Mesoderma cardiogenicum
Septum transversum
Mesoderma capitis
Notochorda

PERIODUS SULCI NEURALIS MATURI ET SOMITORUM IMMATURORUM[11]

Neurulatio
  Plica neuralis
  Plica capitalis
  Plica caudalis
  Plica lateralis corporis
Hiatus umbilicalis
Somiti
Prominentia cardiaca
Sulcus opticus
Placoda otica

PERIODUS TUBI NEURALIS

Neuroporus rostralis
Neuroporus caudalis
Fovea optica
Fovea otica
Primordium cardiacum
Stomatodeum
  Membrana oropharyngealis
Arcus branchialis primus (I)
  Prominentia maxillaris[12]

---

[9] *See* FETAL MEMBRANES, pages E 26 and E 27 for further terminology of extra-embryonic structures.
[10] *Pre-* has been adopted as a prefix to signify a location in front or anterior. *See pro-*, page E 5.
[11] All developmental periods on pages E 9 and E 10 list features of external form only.
[12] Prominence (signifying a local elevation) has been chosen to replace the former inappropriate term, process.

NOMINA EMBRYOLOGICA

Prominentia mandibularis
Sulcus branchialis primus (I)
Arcus branchialis secundus (II)

PERIODUS BRANCHIALIS INITIALIS

Prominentia frontonasalis
Prominentia prosencephalica
  Vesicula optica
Prominentia mesencephalica
Flexura mesencephalica
Prominentia rhombencephalica
Vesicula otica [Otocystus]
Flexura cervicalis
Crista medullae spinalis
Arcus branchiales (I–V)
Sulci branchiales (I–IV)
Prominentia cardiaca
Gemma caudalis

PERIODUS BRANCHIALIS ULTIMA

Placoda nasalis
Placoda lentis
Prominentia hepatis
Prominentia mesonephrica
Gemma membri superioris
Annulus umbilicalis [Anulus]
Cauda

PERIODUS INITIALIS GEMMAE MEMBRORUM

Cupula optica
Fovea lentis
Prominentia frontonasalis[13]
  Prominentia frontalis
  Prominentia nasalis mediana
  Fovea nasalis
  Prominentia nasalis lateralis
Gemma membri
  Membrum superius
  Membrum inferius
  Margo preaxialis
  Margo postaxialis
  Crista ectodermalis apicalis

PERIODUS POSTERIOR GEMMAE MEMBRORUM

Prominentia telencephalica
Prominentia mesencephalica
Prominentia metencephalica
Flexura pontina
Prominentia myelencephalica
Flexura cervicalis
Vesicula lentis
Spiraculum
Tubercula auricularia
Plica opercularis
Sinus cervicalis
Lamina primitiva manus
Lamina primitiva pedis
Tuberculum genitale
Fovea externa cloacalis
Crista mammaria

PERIODUS LABII FISSI

Frons
Nasus
  Naris
Sulcus nasomaxillaris
Premaxilla
Maxilla
Mandibula
Orificium oris
Brachium/Femur
Flexurae membrorum
Antebrachium et Crus primitivum
Manus primitiva et Pes primitivus
  Primordia digitorum
Tuber genitale
Tuberculum genitale
Plica urogenitalis
Sulcus urogenitalis
Proctodeum [Fovea analis]

PERIODUS INITIALIS FETALIS

Plicae palpebrales
Auricula
Digiti
  Digiti primordiales (Inseparati)
  Digiti definitivi (Separati)

---

[13] These are local elevations rather than "processes" in the usual sense of that term.

Plexus venosus cranialis
Phallus
  Sulcus urogenitalis
Labium minus
Labium majus
Tuber scrotale
Raphe anogenitalis
Corpus perineale
Anus

PERIODUS DEFINITIVUS FETALIS

MESODERMA, PER PERIODUM BRANCHIOGENESIS

Mesoderma
  Mesoderma paraxiale
    Somiti[14]
      Dermomyotomi
        Dermatomi[15]
      Myotomi
        prochordales (pro-otici)
        parachordales
        pars epaxialis
        pars hypaxialis
      Sclerotomi[16]
  Mesoderma intermedium[17]
  Mesoderma laterale
    Mesoderma somaticum (parietale)
    Mesoderma splanchnicum (viscerale)
  Mesoderma branchiale
  Mesoderma membrorum
    Massa dorsalis
    Massa ventralis
  Mesenchyma

HISTOGENESIS

ECTODERMA

Epidermis
  Epithelium cuboideum simplex
  Epithelium cuboideum stratificatum
    Periderma
    Epidermis propria
  Epithelium squamosum

  stratificatum
    Cornificatio
    Derivatito
Epithelium tubi neuralis
        (Neuroectoderma)
  Ependymoblasti
  Spongioblasti
    Myelinisatio glioblasti
  Neuroblasti
    apolares
    Processificatio
      Dendrificatio
      Coni augmentales
    bipolares
    unipolares
    multipolares
  Neuroni
Textus cristae neuralis (Ecto-
        mesenchyma)[18]
  Segmenta cristalia
    Ganglia craniopsinalia
  Neuroblasti
  Chromaffinoblasti
  Neurolemmoblasti
    Myelinisatio
  Glioblasti ganglionici
  Melanoblasti
  Chondroblasti
  Odontoblasti
  Mesenchyma capitis
Epithelium sensorium
  Placodae neurales
Epithelium contractile
  Myoepithelium
  Musculi iridis
Epithelium glandulare
Epithelium stomatodeale
  Ameloblasti
  Glandulae salivares
Epithelium proctodeale

MESODERMA

Endothelium
Mesothelium

---

[14] Proposed additions: *Somiti occipitales, cervicales, thoracici, lumbares, sacrales, coccygeales.*
[15] The epidermis and its derivatives are listed in detail on page E 26.
[16] Derivatives listed under Axial Skeleton, page E 13.
[17] Derivatives listed under Urogenital System, page E 18.
[18] Some cells of neural crest origin form structures (such as branchial arches and dentin) similar to parts typically mesodermal in origin. The term *ectomesenchyma* has also been proposed.

Epithelium mesodermale
  Epithelium glandulare
Textus epitheliodeus[19]
Mesenchyma
  Angioblasti
  Textus myeloideus
    Haemocytoblasti [Hemo-]
  Textus lymphoideus
    Lymphoblasti
  Fibroblasti
    Fibrillogenesis
  Lipoblasti
  Chondroblasti
  Osteoblasti
    Substantia osteoidea
    Osteoclasti[20]
  Odontoblasti[21]
  Myoblasti
    Myofibrillogenesis
    Status mononuclearis
      Musculus nonstriatus
      Musculus cardiacus
      Musculus skeletalis
    Status multinuclearis
      Musculus skeletalis
        Myotubuli
    Textus conducens
      Fibrae conductens (F. Purkinjii)
      Nodi
  Chorda nephrogenica
    Tubuli renales
    Epithelium transitionale

ENDODERMA

Epithelium
  simplex
    squamosum
    cuboidale
    columnare
  pseudostratificatum
  stratificatum
    squamosum
      cornificatum
    transitionale
  ciliatum
  glandulare
  sensorium

Cellulae germinales primordiales
Angioblasti[22]

ORGANOGENESIS

SYSTEMA SKELETALE

Skeletogenesis primarius
  Chordagenesis
  Chondrogenesis
    Mesoderma blastemale
      Centrum chondrificationis
    Precartilago
      Perichondrium
        Stratum chondrogeneticum
      Cartilago embryonica
        Status proliferans
          Incrementum appostionale
          Incrementum interstitionale
        Typus hypertrophicus
        Subtypi differentes
Osteogenesis
  Osteogenesis membranacea
    Membrana cellulare
    Os spongiosum [O. trabeculare]
    Periosteum
      Stratum osteogeneticum
        Os spongiosum/compactum
  Osteogenesis cartilaginea
  Ossificatio perichondralis
    Perichondrium
      Stratum osteogeneticum
        Os periochondriale
          Annulus [Anulus] osseus
  Ossificatio endochondralis
    Cartilago calcificata
    Gemma osteogenetica primaria
    Canalis eruptivus
    Centrum ossificationis primarium
      (Centrum diaphysiale)
    Zonae differentiationae
    Os primarium
    Cavitas medullaris primaria
    Gemma osteogenetica secundaria
    Centrum ossificationis secundarium
      (Centrum epiphysiale)

[19] Suprarenal cortex; gonadal parenchyma.
[20] Contributions from osteocytes and mesenchyme cells are also recognized.
[21] Experimental evidence implicates the neural crest as the source of these cells.
[22] The germ-layer origin of this tissue in some animals and regions has been interpreted as endodermal.

Os secundarium
  Os intertextum (O. prenatale)
    Os spongiosum [O. trabeculare][23]
  Os compactum immaturum
    Osteona primaria[24]
  Os compactum definitivum (postnatale)
    Lamellae osseae
    Osteona secundaria[24]
  Medulla ossis
Skeleton axiale
  Notochorda (Chorda dorsalis)
    Vagina notochordalis
    Nuclei pulposi
  Sclerotomi
    Pars rostralis
      Vertebrae
        Precartilaginea
        Cartilagines
        Osseae
          Centrum
          Arcus vertebralis
          Processus spinosus
      Costae
        precartilagineae
        cartilagineae
        osseae
    Pars caudalis
      Discus intervertebralis
  Mesoderma sternalis
    Cartilago episternalis
    Cartilago sternalis
      Sternebrae
      Processus xiphoideus
Cranium
  Desmocranium
  Chondrocranium
    Viscerocranium
  Osteocranium
Neurocranium
  Meninx primitiva
    Meninges
  Calvaria
  Sclerotomi occipitales
    Cartilago parachordalis
    Cartilago occipitalis

Cartilago hypophysialis
Cartilago trabecularis
  Cartilago sphenoidalis
Capsula nasalis
Capsula otica
Fonticuli
Viscerocranium [Arcus branchiales]
  Cartilago branchialis
    Arcus primus (I)
      Pars dorsalis[25]
        Maxilla (plerumque)
      Pars ventralis[26]
        Mandibula (partim)
    Arcus secundus (II)
      Pars dorsalis[27]
      Pars ventralis
    Arcus tertius-sextus (III–VI)
      Pars copularis
    Larynx
Skeleton appendiculare
  Blastema skeletale
  Cartilagines
    Lamina epiphysealis
    Metaphysis
    Diaphysis
  Osseae
  Articulationes
    Cartilago articularis
    Ligamenta primordialia
    Synovialia primordialia
    Capsulae articulares

SYSTEMA MUSCULARE

Myogenesis[28]
Mesoderma prechordale et parachordale
  Myotomi
    Myotomi prechordales [pre-otici]
      Primordium musculare oculi
    Myotomi occipitales
      Primordium musculare linguae
    Myotomi spinales
      Pars epaxialis
        Primordium musculi extensoris
        spinae
      Pars hypaxialis

---

[23] Since trabeculae (beams) are rare, it has been suggested that Os muraliosum (wall) would be a more appropriate term.
[24] Also called Haversian systems.
[25] Also called quadrate cartilage (*Cartilago quadrata*).
[26] Also called Meckel's cartilage (*Cartilago Meckeliensis*).
[27] Also called Reichert's cartilage (*Cartilago Reicherti*).
[28] *See also* under HISTOGENESIS, page E 11, 12.

Regio cervicalis
  Primordium musculare sterno-
    cleido-trapezii (partim)
  Primordia muscularia
    Primordium musculorum
      geniohyoideum
    Primordium musculorum
      infrahyoidorum
    Primordium musculorum
      prevertebralium
    Primordium musculorum
      scalenorum
    Primordium pectorale
      gemmae membri (partim)
  Lamina diaphragmatica
    thoracica (plerumque)
Regio thoracolumbalis
  Primordium musculorum
    flexorum spinae
Regio sacrococcygealis
  Lamina diphragmatica pelvica
Musculi unisegmentales
Musculi multisegmentales
Mesoderma intermedium
  Musculi nonstriati ductuum
    urogenitalium
Mesoderma laminae lateralis
  Mesoderma branchiomericum
    Primordium musculare arcus primi (I)
      Musculi masticatorii (et cetera)
    Primordium musculare arcus secundi
      (II)
      Musculi faciales (et cetera)
    Primordia muscularia arcuum tertii
      quarti, et sexti (III, IV &
      VI)[29]
    Primordium musculare palatale
    Primordium musculare pharyngeale
    Primordium musculare sterno-
      cleido-mastoidei et trapezii
      (partim)
  Mesoderma somaticum
    Mesenchyma gemmae membri
    Primordium musculare dorsale
    Primordium musculare ventrale
  Diphragma (partim)
  Sphincter cloacalis (plerumque)[30]

Primordium sphincteris ani (partim)
Primordium sphincteris urogenitalis
Mesoderma splanchnicum
  Musculi canalis alimentarii
  Musculi canalis respiratorii
  Musculi cardiaci
  Musculi vasculares
  Musculi vesico-urethrales

SYSTEMA DIGESTIVUM

Primordia
  Saccus vitellinus primitivus
    Saccus vitellinus proximalis
    Saccus vitellinus distalis
  Lamina prochordalis
  Stomatodeum
  Pre-enteron
  Mesenteron
  Metenteron
  Proctodeum

CAVITAS ORIS

Stomatodeum [Stomodeum]
Saccus hypophysialis[31]
  Adenohypophysis [Lobus anterior
    hypophysis]
  Pars distalis
  Pars tuberalis
  Lumen residuale
  Pars intermedia
  Pars pharyngealis
Prominentia maxillaris[32]
Prominentia nasalis mediana
Prominentia mandibularis
Membrana oropharyngealis [M. bucco-
  pharyngealis]
Primordia lingualia
  Tuberculum linguale distale
  Gemma lingualis media
  Sulcus terminalis
  Tuberculum linguale proximale
  Copula
  Gemmae gustatoriae

---

[29] The last arch becomes the sixth in the series if a controversial fifth is recognized as such.
[30] Derivatives forming striated muscle occur in both the anal and urogenital regions.
[31] Also called Rathke's pouch.
[32] *See* footnote 12, page E 9.

Sulcus linguogingivalis
  Gemma glandulae submandibularis[33]
  Gemmae glandulae sublingualis
Processus palatinus medianus
  Premaxilla (Palatum primum)[34]
  Foramen incisivum
Processus palantinus lateralis
  Palatum proprium
Taenia [tenia] labiogingivalis
  Sulcus labiogingivalis
    Vestibulum
    Gemma glandulae parotideae
  Labii oris
  Bucca
  Gingivae

Dens
  Lamina dentalium
    Relicti laminales[35]
    Organa enamelaria
      Status gemmalis
      Status cappalis
      Status campanalis
    Epithelium enamelum[36] externum
    Reticulum enamelum
    Epithelium enamelum internum
      Ameloblasti
        Prismae enamelariae
        Lamina basalis enameli[37]
    Vagina radicalis epithelialis
      Diaphragma vaginae radicis
      Porus vaginae radicis
    Cuticula dentalis
  Papilla dentalis
    Pulpa dentalis
    Odontoblasti
      Predentinum
      Dentinum
  Saccus dentalis
    Lamina cementoblastica
      Cementum
    Lamina periodontoblastica

Lamina osteoblastica
Alveolus dentalis
Canalis eruptivus
Dens deciduus
Dens permanens

PRE-ENTERON

Pharynx primitiva
  Arcus branchiales[38]
  Sacci pharyngeales
  Saccus primus (I)
    Recessus tubotympanicus
      Tuba auditiva [auditoria]
      Cavitas tympanica
      Cellulae tympanicae
      Antrum mastoideum
        Cellulae mastoideae
  Saccus secundus (II)
    Fossa supratonsillaris
  Saccus tertius (III)
    Pars dorsalis
      Gemma parathyroidea inferior
      [caudalis]
    Pars ventralis
      Gemma thymica major
  Saccus quartus (IV)
    Pars dorsalis
      Gemma parathyroidea superior
      [rostralis]
    Pars ventralis
      Gemma thymica minor
  Saccus quintus (V)
    Corpus ultimobranchiale[39]
Diverticulum thyroideum
  Foramen caecum [cecum]
  Ductus thyroglossalis
  Glandula thyroidea
Oesophagus primitivus [Eso-]
Gaster [Ventriculus] primitivus
Duodenum primitivum
Diverticulum hepaticum

[33] Named the mandibular gland in domestic mammals.
[34] According to some authorities there is no true premaxilla in human development. The term *palatum primum* would therefore be a possible substitute.
[35] Persistent remnants of the involuting lamina may give rise to masses of dental tissue within the gum, or to cysts within the alveolus.
[36] *Adamantinum* is an older and less easily recognized synonym for "enamel."
[37] This basement membrane, separating early ameloblasts and odontoblasts, has commonly borne the inappropriate name of *membrana preformativa*.
[38] *See* pages E 10, E 13 and E 14.
[39] Some investigators question the origin of these structures from rudimentary fifth branchial sacs.

NOMINA EMBRYOLOGICA

Ductus hepatopancreaticus
  Antrum hepaticum
  Gemmae pancreaticae ventrales
    Gemma dextra
    Gemma sinistra
  Ductus pancreaticus ventralis
    Pancreas ventrale
    Systema ductale primitivum
    Acini
    Insulae
  Ductus choledochus [biliaris]
    Ductus hepatici
      Lamina hepaticae[40]
    Ductus cysticus
      Vesica biliaris [fellea]
Gemma pancreatica dorsalis
  Pancreas dorsale
    Ductus pancreaticus dorsalis
    Systema ductale primitivum
      Acini
      Insulae
Anastomosis ductalis
Duodenum (partim)

MESENTERON

Ansa umbilicalis intestini
  Crus cranialis
  Crus caudalis
  Rotatio ansae intestinalis
  Duodenum (partim)
  Jejunum
  Ileum
    Pedunculus [ductus] vitellinus
      Vestigium pedunculi vitellini[41]
  Bulla caecale [cecale]
    Caecum [cecum]
    Appendix vermiformis
  Colon ascendens
  Colon transversum (partim)

METENTERON

Colon transversum (partim)
Colon descendens
Colon sigmoideum

Cloaca
  Rectum
  Urenteron

PROCTODEUM

Membrana analis
Canalis analis
Anus

SYSTEMA RESPIRATORIUM

NASUS

Placoda nasalis
  Placoda olfactoria
Fovea nasalis
Saccus nasalis
  Epithelium olfactorium
    Epithelium vomeronasale
  Epithelium respiratorium
Prominentia frontonasalis[42]
Prominentia nasalis lateralis
Prominentia nasalis mediana
  Regio premaxillaris
    Palatum primitivum
    Membrana oronasalis
      Choana primitiva
  Prominentia septalis
    Septum nasale
Naris externa
Naris interna (Choana)
Rugae turbinatae
Sulci sinus paranasalis

ARBOR RESPIRATORIUM

Eminentia hypobranchialis
  Epiglottis
Sulcus laryngotrachealis
  Crista tracheo-oesophagealis [-eso-][43]
  Crista oesophagealis lateralis

---

[40] The traditionally named "hepatic cords" are actually thin, anastomosing plates.
[41] The commonest form of this remnant is known as Meckel's diverticulum of the ileum.
[42] See footnote 12, and pages E 9 and E 10.
[43] The Latin stem is really *oeso-*; the "e" in *ēso-* is therefore long in pronunciation.

Septum tracheo-oesophageale
Gemma pulmonaria
Tubus laryngotrachealis
  Glottis primitiva
  Tuber arytenoideum
  Larynx
  Trachea
  Saccus pulmonarius primitivus
    Bronchus primarius
    Gemmae lobales
    Gemmae bronchopulmonariae
Pulmo fetalis
  Periodus pseudoglandularis
    Lobi
    Bronchi
  Periodus canalicularis
    Bronchioli
      Bronchioli respiratorii
  Periodus alveolaris
    Ductuli alveolares
    Sacculi alveolares
      Gemmae alveolares

COELOMATA ET SEPTA

Coeloma extra-embryonicum
  Cavitas chorionica
  Coeloma umbilicale
Coeloma intra-embryonicum
  Vesiculae coelomaticae
  Primordium pericardiale
    Cavitas pericardialis
    Canalis pericardioperitonealis
Septum transversum[44]
  Hiatus pleuropericardialis
    Plica pleuropericardialis
      Membrana pleuropericardialis
  Hiatus pleuroperitonealis
    Plica pleuroperitonealis
      Membrana pleuroperitonealis[45]
  Cavitas pleuralis
  Diaphragma[46]

Cavitas peritonealis
  Recessus pneumato-entericus
    Bursa infracardiaca
    Vestibulum
    Plica gastropancreatica
    Bursa omentalis
  Sinus coelomaticum
  Spatium subphrenicum
  Saccus vaginalis
    Cavitas scrotalis
    Saccus inguinalis[47]
Hiatus umbilicalis
  Annulus umbilicalis [Anulus]

MESENTERIA ET PLICAE PERITONEALES[48]

Mesenterium dorsale primitivum
  Meso-oesophageum [-eso] dorsale
  Mesogastrium dorsale
    Omentum majus
      Plica phrenicosplenica
      Plica gastrophrenica
      Plica gastrosplenica
      Plica gastrocolica
      Plica phrenicocolica
  Mesoduodenum dorsale
    Fascia retinens rostralis[49]
  Mesenterium dorsale commune
    Mesojejunum
    Meso-ileum
    Mesocolon
    Mesorectum
Mesenterium ventrale primitivum
  Meso-oesophageum [-eso] ventrale
  Mesogastrium ventrale
    Omentum minus
      Plica hepatogastrica
      Plica hepatoduodenalis (partim)
    Plica falciformis
    Plica coronaria
    Plica triangularis

---

[44] Primordium of the diaphragm, in part.
[45] Primordium of the diaphragm, in part.
[46] Other contributions come from mesenchyme of the body wall, meso-oesophagus, mesenchyme behind the suprarenals, mesonephros and gonads, and alongside the aorta.
[47] This rudimentary equivalent of the scrotal cavity in the female is also known as the canal of Nuck.
[48] These supposed supports are customarily called ligaments, but peritoneal folds (*plicae*) is more appropriate.
[49] Retention bands form within the primitive duodenal mesentery, thus regulating the movement and fixation of the duodenal loop. Part of them becomes muscular and then is known as the suspensory muscle of the duodenum (formerly named the ligament or muscle of Treitz).

NOMINA EMBRYOLOGICA

Mesoduodenum ventrale
    Plica hepatoduodenalis (partim)
  Mesocystis[50]
    Plica umbilicalis mediana
Plica umbilicalis medialis
Mesenterium urogenitale
  Plica suspensoria gonadalis
  Mesorchium
  Mesovarium
Septum urogenitale
Mesenterium ductus paramesonephrici
  Plica lata uterina
    Mesosalpinx
    Mesometrium
Plica gubernacularis
Plica inguinalis
Mesenchyma gubernaculare
  Plica ovarii propria
  Plica teres uteri
  Gubernaculum testis/ovarii
    Descensus testis

Pars colligens
Plica mesonephrica

METANEPHROS
Diverticulum metanephricum
  Ureter
  Pelvis renalis
  Calix renalis
  Ductus papillares
  Tubuli colligentes

BLASTEMA METANEPHROGENICUM
Capsula renalis
Nephroni
  Corpusculum renale
    Capsula glomerularis
    Glomerulus
  Tubulus secretorius
    Tubulus convolutus
    Ansa nephronis
Tubuli uriniferentes (plerumque)

SYSTEMA UROGENITALE

SYSTEMA RENALE

Mesoderma intermedium
  Nephrotomi

PRONEPHROS[51]
Tubuli pronephrici
  Nephrostoma
  Canaliculus nephrostomaticus
  Glomerulus coelomaticus
Ductus pronephricus

MESONEPHROS
Crista mesonephrica
Ductus mesonephricus
Corpusculi mesonephrici
  Capsula glomeruli
  Glomerulus
Tubuli mesonephrici
  Pars secretoria

SYSTEMA GENITALE

GONADA
Status indifferens
  Crista gonadalis
    Epithelium superficiale
    Mesenchyma
  Cellulae germinales primordiales
    Chordae gonadales[52]
      Cellulae germinales[53]

OVARIUM
Epithelium superficiale[54]
Chordae gonadales
Medulla
  Chordae medullares[52]
    Folliculi medullares
  Rete ovarianum
Cortex
  Chordae corticales
    Ovogonia nuda[55]
    Racemus ovorum
    Folliculi corticales

[50] A transitory ventral mesentery that obliterates as the urinary bladder fuses with the ventral body wall.
[51] It is questionable whether in mammals any of the structures listed as subheadings under "pronephros" ever differentiate beyond mere primordia.
[52] At first there are anastomosing plates that secondarily break up into cords (*chordae*).
[53] The cytogenesis and morphology of germ cells are treated on pages E 6 and E 7.
[54] The traditional name, germinal epithelium, is misleading and inappropriate.
[55] For details on the cytogenesis and morphology of these cells, *see* pages E 6 and E 7.

primordiales
primarii
   Cellulae folliculares
secondarii
   Antrum
atretici
Corpus luteum
Corpus albicans
Stroma
   Cellulae (glandulae) interstitiales[56]
   Tunica albuginea

TESTIS
  Epithelium superficiale
  Chordae gonadales
    Tubuli seminiferi
     Tubuli contorti
      Cellulae germinales[57]
      Cellulae sustentaculares
     Tubuli recti
    Rete testis
  Stroma
    Tunica albuginea
    Mediastinum testis
    Septula testis
    Cellulae interstitiales

DUCTUS GENITALES
  Tubuli mesonephrici
    Epoöphoron
    Paroöphoron
    Ductuli efferentes
    Ductuli aberrantes rostrales
    Ductuli aberrantes caudales
    Paradidymis
  Ductus mesonephricus
    Appendix vesicularis
    Ductus epoöphorontis
     Ductus deferens vestigialis[58]
    Ductus epididymidis
     Appendix epididymidis
    Ductus deferens
     Ampulla ductus deferentis

Glandula seminalis[59]
Ductus ejaculatorius
Trigonum vesicae[60]

CRISTA GONADALIS
  Sulcus paramesonephricus
  Ductus paramesonephricus[61]
    Pars pre-infundibularis
     Appendix vesicularis
     Appendix testis
    Pars infundibularis
    Pars postinfundibularis
     Tuba uterina
     Primordium uterovaginale
     Uterus
     Vagina (partim)
     Vagina masculina (partim)[62]

CLOACA

Membrana cloacalis
Septum urorectale
Rectum
Sinus urogenitalis primitivus
  Canalis vesico-urethralis
    Pars vesicalis
    Pars urethralis
  Tuberculum sinuale
    Bulbus sinuvaginalis
     Vagina (partim)
     Hymen
    Bulbus sinu-utricularis
     Vagina masculina (partim)[62]
Sinus urogenitalis definitivus
  Pars vesicalis
    Urachus
    Plica umbilicalis media
  Pars pelvica
    Urethra feminina/masculina
    (partim)
    Pars prostatica urethrae
    Gemmae glandulares prostaticae ιe

---

[56] These occur in the fetal ovaries of most mammals.
[57] For details on the cytogenesis and morphology of these cells, *see* pages E 6 and E 7.
[58] Residual remnants, especially at the caudal end, have been called Gartner's duct.
[59] Since in man this gland does not primarily or normally store constituents of the semen, the term "seminal vesicle" is not appropriate.
[60] Some authorities believe that the original mesodermal lining is replaced by endoderm.
[61] The original Subcommittee invented *Ductus feminus* as a more significant term, but in revision a majority preferred to conform with *Nomina Anatomica* usage.
[62] In gross anatomy, called the prostatic utricle.

Pars membranacea urethrae
Pars phallica
Sulcus urethralis
Pars spongiosa urethrae
Bulbus urethralis
Vestibulum vaginale (partim)
Glandula bulbo-urethralis
Glandula vestibularis major
Proctodeum
Membrana analis
Canalis analis
Anus
Prominentia perinealis
Membrana urogenitalis
Ostium urogenitale
Vestibulum vaginale (partim)

GENITALIA EXTERNA
Tuberculum genitale
Phallus primitivus
Pars dorsalis penis
Pars dorsalis clitoridis
Sulcus coronarius
Glans penis
Glans clitoridis
Lamella glandularis
Fossa navicularis urethrae
Lamella glandopreputialis
Plicae urogenitales
Labium minus[63]
Pars ventralis penis
Sulcus urogenitalis definitivus
Vestibulum vaginale (plerumque)
Urethra primitiva
Tubercula labioscrotalia
Labium majus[63]
Commissura caudalis
Scrotum
Raphe scrotalis

SYSTEMA CARDIOVASCULARE

COR

Mesoderma splanchnicum
Mesoderma cardiogenicum
Primordium endocardiale
Primordium epimyocardiale

Cor primordiale
Primordium sinus venosi
Primordium atriale
Primordium ventriculare endocardiale
Ventriculus saccularis primitivus
Cor tubulare simplex
Atrium primitivum
Junctio atrioventricularis
Ventriculus primitivus
Bulbus cordis primitivus
Endocardium primitivum
Cardioglia
Myocardium primitivum
Epicardium primitivum
Mesocardia
Cor sigmoideum
Sinus venosus
Pars transversa
Cornu (dextrum/sinistrum)
Ostium sinuatriale
Valvula sinuatrialis
Atrium primitivum
Canalis atrioventricularis
Tuber endocardiale atrioventriculare
Ventriculus primitivus
Ansa bulboventricularis
Sulcus bulboventricularis
Ostium bulboventriculare
Tuber endocardiale
Valva bulboventricularis
Bulbus cordis
Crista bulbaris
Septum spirale
Cor quadricameratum
Sulcus bulboventricularis
Conus arteriosus
Sulcus interventricularis
Sulcus interatrialis
Sulcus coronarius
Sinus venosus
Sinus coronarius (partim)
Vena obliqua (partim)
Tuberculum intervenosum
Valva sinus venosi
Septum spurium
Crista terminalis
Valva venae cavae inferioris
Valva sinus coronarii
Atrium primitivum
Septum primum

---

[63] In domesticated mammals these *labia* are fused, comprising a *labium vulvae*.

E 20

Foramen (interatriale) primum
Foramen (interatriale) secundum
Septum secundum
Foramen ovale
Limbus fossae ovalis
Atrium dextrum/sinistrum
Canalis atrioventricularis
Tuber endocardiale
Valva atrioventricularis
Valva mitralis (bicuspidalis)
Valva tricuspidalis
Ventriculus primitivus
Septum interventriculare
Foramen interventriculare
Pars muscularis
Pars membranacea
Trabeculae carneae
Ventriculus dexter/sinister
Valva semilunaris
Valva aortica [V. aortae]
Valva pulmonalis

SYSTEMA VASCULARE

Mesenchyma
Textus angioblasticus
Insulae sanguineae
Endothelioblasti
Haemocytoblasti [Hemo-]
Rete capillare primitivum
Circulatio embryonica
Rete vasculare
Phasis bilateralis
Phasis inequalis

ARTERIAE
Saccus aorticus
Truncus arteriosus
Crista aorticopulmonalis
Septum aorticopulmonale
Aorta ascendens
Valva aortica
Truncus pulmonalis
Valva [trunci] pulmonalis

A. carotis communis (partim)
Arcus aorticus primus (I)
Arcus aorticus secundus (II)
Arcus aorticus tertius (III)
A. carotis communis (partim)
A. carotis interna (partim)
A. carotis externa
Arcus aorticus quartus (IV)
Arcus aortae definitivus
Truncus brachiocephalicus
A. subclavia dextra (partim)
Arcus aorticus quintus (V)[64]
Arcus aorticus sextus (VI)
A. pulmonalis[65]
Ductus arteriosus
Ligamentum arteriosum
Aorta dorsalis
A. carotis interna (plerumque)
A. subclavia dextra (partim)
Aa. intersegmentales dorsales
Rami dorsales
Anastomoses dorsales
A. vertebralis
Anastomosis vertebralis
A. basilaris[66]
Anastomoses ventrales
Truncus thyrocervicalis
Truncus costocervicalis
Rami ventrales
A. subclavia
dextra (partim)
sinistra
A. axialis membri superioris
Aa. intercostales
Aa. lumbales [lumbares]
Aa. segmentales laterales
A. phrenica
A. suprarenalis media
A. renalis
A. gonadalis
Aa. segmentales ventrales[67]
Aa. vitellinae
Truncus coeliacus
Aa. mesentericae
A. allantoica
A. umbilicalis

---

[64] This arch is controversial as a legitimate member in the series of aortic arches.
[65] These arteries arise from the aortic sac and grow to the lung bud, whereas the sixth aortic arches sprout from the aorta and join them secondarily.
[66] This artery is said to result from the consolidation of two independent channels ventral to the brain, whereas connection with the vertebral arteries is secondary.
[67] These branches are irregularly and imperfectly segmental.

Plica umbilicalis medialis[68]
A. iliaca interna
A. axialis membri inferioris
A. sacralis mediana

Venae
Vv. extra-embryonicae
V. vitellina
V. allantoica
V. umbilicalis
Vv. intra-embryonicae
V. umbilicalis
Ligamentum teres hepatis
Plexus venosus visceralis
Vv. viscerales
V. pulmonalis communis
Vv. vitellinae
V. portae hepatis [portalis]
Vv. afferentes hepatis
Ductus venosus
Ligamentum venosum
Vv. efferentes hepatis [Venae hepaticae]
Pars hepatica venae cavae inferioris[69]
Vv. somaticae
V. precardinalis
V. capitis primaria
V. jugularis interna
V. jugularis externa
Anastomosis precardinalis
Truncus brachiocephalicus
V. subclavia
Vena cava superior (partim)
V. cardinalis communis
Vena cava superior (partim)
Sinus coronarius (partim)
V. obliqua (plerumque)
V. postcardinalis
V. azygos (partim)
Vena cava inferior (partim)
V. subcardinalis
Anastomosis subcardinalis
Vena cava inferior (partim)
V. renalis sinistra
V. supracardinalis
V. azygos (plerumque)

Vena cava inferior (partim)
Anastomosis sub-supracardinalis[69]
V. gonadalis
Vv. intersegmentales
Vv. marginales membrorum
Vv. membri superioris
Vv. membri inferioris

SYSTEMA LYMPHATICUM

Mesenchyma
Textus lymphoblasticus
Sacci lymphatici
Saccus jugularis
Saccus subclavius
Cisterna chyli
Saccus retroperitonealis
Saccus iliacus
Capillares lymphatici
Vasa lymphatica
Ductus thoracicus duplicatus (dexter/ sinister)
Ductus thoracicus definitivus
Junctio lymphaticovenosa
Primordia nodorum lymphaticorum
Tonsillae
Primordia splenica [P. lienis]
Splen [Lien]
Thymus[70]

SYSTEMA NERVOSUM

Neurogenesis[71]
Lamina neuralis
Plica neuralis
Sulcus neuralis
Crista neuralis

Tubus neuralis
Canalis neuralis
Stratum ependymale
Stratum palliale
Stratum marginale
Lamina dorsalis
Epithelium plexus choroidei

---

[68] For decades the *Nomina Anatomica* term was Lig. umbilicale laterale.
[69] The several components of the *inferior vena cava* present difficulties of comparison in different animals types.
[70] *See* third and fourth pharyngeal pouches, page E 15.
[71] *See also* HISTOGENESIS, page E 11.

Lamina dorsolateralis
Sulcus limitans
Lamina ventrolateralis
Lamina ventralis
Neuroporus
   Neuroporus rostralis [cranialis]
   Neuroporus caudalis
Lamina terminalis primitiva

ENCEPHALON
  Substantia alba
  Substantia grisea
  Liquor cerebrospinalis
  Vesiculae encephalicae

PROSENCEPHALON
  Cavitas prosencephalica

Rhinencephalon
  Cavitas rhinencephalica
  Bulbus olfactorius
  Cortex piriformis
  Fissura rhinalis
  Area paraterminalis
  Hippocampus primitivus
    Hippocampus
    Gyrus dentatus
    Vestigium hippocampale
    Systema fornicale

Telencephalon
  Cavitas telencephalica
  Pars mediana
    Lamina terminalis definitiva
    Lamina commissuralis
      Commissura anterior
      Commissura hippocampalis
      Commissura neopallialis
      Ventriculus tertius (partim)
    Hemispherium cerebrale
      Ventriculus lateralis (dexter/sinister)
      Foramen interventriculare
      Stratum choroideum epitheliale
      Tela choroidea
        Fissura choroidea
      Velum interpositum
      Pars striata hemispherii
        Vestigium striatum mediale
        Vestigium striatum laterale
        Globus pallidus

Palaeocortex [Paleocortex]
Pars suprastriata hemispherii
Hippocampus primitivus
Neocortex
    Cortex trilaminaris primarius
    Cortex stratificatus definitivus

Diencephalon
  Cavitas diencephalica
    Ventriculus tertius (partim)
  Tela choroidea
  Gemma pinealis
  Gemma neurohypophysialis

MESENCEPHALON
  Cavitas mesencephalica
    Aquae ductus mesencephalicus [cere-
    bri]
  Flexura cephalica

RHOMBENCEPHALON
  Cavitas rhombencephalica
    Ventriculus quartus
  Tela choroidea
  Rhombomeri

Metencephalon
  Flexura pontina
  Labium rhombencephalicum
    Primordium cerebellare
  Extensio bulbopontina

Myelencephalon [Medulla oblongata]
  Flexura cervicalis

MEDULLA SPINALIS
  Canalis centralis
  Septum medianum dorsale
  Fissura mediana ventralis
  Filum terminale
  Intumescentia cervicalis
  Intumescentia lumbosacralis
    Conus medullaris
  Crista neuralis
    Segmenta cristalia
      Ganglia craniospinalia
        [encephalospinalia]
      Ganglia autonomica[72]
        Cellulae chromaffinae
          Medulla suprarenalis
  Placodae neurales

---

[72]   An added contribution from the neural tube is advocated by some investigators.

E 23

Placodae epibranchiales
Placodae suprabranchiales
Placodae dorsolaterales
Nervi craniospinales [encephalospinales]

MENINGES

Mesenchyma
  Meninx primitiva
  Ectomeninx
    Lamina interna periostealis
    Dura mater craniospinalis
  Endomeninx
    Arachnoidea Mater craniospinalis[73]
      Reticulum arachnoideum
    Pia mater craniospinalis[73]
    Tela choroidea

SYSTEMA SENSORIUM

ORGANUM GUSTUS [GUSTATO-RIUM] (*see* page E 14)

ORGANUM OLFACTUS [OLFACTO-RIUM] (*See* page E 16)

OCULUS

Prosencephalon
  Recessus opticus
Vesicula optica
  Cavitas optica
  Pedunculus opticus
Cupula optica
  Labrum cupulae
  Lamina externa cupulae
  Spatium intraretinale
  Lamina interna cupulae
  Cavitas cupularis
  Fissura optica
Placoda lentis
  Fovea lentis
  Porus lentis

Vesicula lentis
  Cavitas lentis
  Epithelium lentis superficiale
  Epithelium lentis profundum
    Fibrae lentis
    Capsula lentis

Retina

Lamina interna cupulae
  Pars optica retinae
    Stratum ependymale
    Stratum photosensorium [S. sensile]
  Stratum palliale
    Stratum nucleare internum [S. bipolare]
    Stratum ganglionare
  Stratum marginale
    Stratum neurofibrarum
  Ora serrata
  Pars caeca [ceca] retinae
    Pars ciliaris retinae
      Epithelium ciliare
    Pars iridica retinae
      Epithelium iridicum

Lamina externa cupulae
  Stratum pigmentosum
    Musculus sphincter pupillae[74]
    Musculus dilator pupillae[74]

Mesenchyma opticum
  Tunica vascularis lentis
  Mesenchyma camerae vitreae
    Arteria lentis
      Arteria hyaloidea
      Canalis hyaloideus
    Corpus vitreum
      Membrana vitrea
  Mesenchyma camerae aquosae[75]
  Epithelium camerae aquosae[75]
  Humor aquosus
  Mesenchyma capsulare
    Tunica interna[76]
      Tunica vasculosa
        Lamina vasculosa
        Lamina pigmentosa

---

[73] An additional contribution from the neural crest is advocated by some experimental embryologists.
[74] This muscle is said to be a local modification of the pigment epithelium.
[75] The *camera aquosa* = *camerae anterior et posterior* of *Nomina Anatomica* = "anterior segment" in some clinical circles.
[76] Corresponding to the *endomeninx* of the brain.

Musculus ciliaris
Stroma iridis
Membrana pupillaris
Tunica externa [T. fibrosa][77]
Sclera
Cornea

ORGANA OCULI ACCESSORIA

Plicae palpebralis
Palpebrae
Epithelium ectodermale
Epidermis
Cilia
Epithelium conjunctivale
Gemmae glandulae palpebrales
Ductus lacrimalis
Gemmae glandulae lacrimales
Crista nasolacrimalis
Saccus nasolacrimalis
Ductus nasolacrimalis
Canaliculi lacrimales
Phasis conjunctionis palpebrarum
Tunica conjunctiva pelpebrarum
Tunica conjunctiva bulbi [bulbaris]
Epithelium corneae [cornealis]

AURIS

AURIS INTERNA
Placoda otica
Fovea otica
Vesicula otica [Otocystis]
Labyrinthus membranaceus
Saccus vestibularis
Laminae semicirculares
Foci absorptionis
Ductus semicirculares
Utriculus
Sacculus
Ductus reuniens
Saccus cochlearis

Ductus cochlearis
Lagaena [Lagena]
Organum spirale[78]
Diverticulum endolymphaticum
Ductus endolymphaticus
Saccus endolymphaticus
Capsula otica
Mesenchyma oticum
Labyrinthus cartilaginosus
Labyrinthus osseus
Canales semicirculares
Vestibulum
Cochlea

AURIS MEDIA
Saccus pharyngealis primus (I)
Recessus tubotympanicus[79]
Tuba auditiva
Cavitas tympanica
Cellulae tympanicae
Antrum mastoideum
Cellulae mastoideae
Membrana branchialis prima (I)
Membrana tympanica[80]
Arcus branchialis primus (I)
Cartilago dorsalis[81]
Incus (plerumque)[82]
Cartilago ventralis[83]
Malleus (plerumque)[82]
Musculus tensor tympani
Arcus branchialis secundus (II)
Cartilago dorsalis[84]
Stapes (partim)[82]
Musculus stapedius

AURIS EXTERNA
Sulcus branchialis primus (I)
Meatus acusticus externus
Arcus branchiales primus et secundus (I, II)
Colliculi aurales
Pinna [Auricula]

[77] Corresponding to the dura mater of the brain.
[78] The "Organ of Corti." An eponymous variant suggested is *Organum cortiense*.
[79] The second pharyngeal pouch may also contribute.
[80] The final membrane is more extensive than the primary association of the first branchial groove and first pharyngeal pouch.
[81] Also known as Meckel's cartilage.
[82] Modern studies have modified the exclusive arch origin of these ossicles.
[83] Also known as the quadrate cartillage (*Cartilago quadrata*).
[84] Also known as Reichert's cartilage (*Cartilago reichertii*).

E 25

SYSTEMA INTEGUMENTALE

Ectoderma
  Epidermis
    Periderma
    Stratum intermedium
    Stratum basale
    Epidermis definitiva
  Ungues[85]
    Campus unguis
    Plica unguis
    Matrix unguis
    Lamina unguis
  Eponychium
  Hyponychium
  Pili
    Gemma pili
      Truncus [Scapus] pili
      Folliculus epithelialis
    Lanugo
    Vellus
  Glandulae
    Glandulae sudoriferae
    Glandulae sebaceae
    Crista mammaria
      Gemma primaria
        Ductus
        Alveoli
      Glandula mammaria
      Fovea mammaria
      Papilla mammaria
      Glandulae areolares[86]
  Vernix caseosa
Melanoblasti
Mesenchyma
  Dermis [Corium][87]
    Papillae
  Vagina dermalis pili
  Papilla pili
  Basis unguis
  Stroma glandulae
  Tela subcutanea

MEMBRANAE FETALES MAMMALIUM

MODI ET CURSUS PROGRESSUS

Distributio in utero
Implantatio (Nidatio)
  Definitio situs
  Orientatio
  Affixio
    centralis
    eccentrica
      mesometrialis
      antimesometrialis
      orthomesometrialis
    superficialis
    interstitialis
  Radicula affixiva (Bulla affixiva; Conus affixivus)
Amniogenesis
  Plicatio
  Cavitatio
Entypia
  Inversio sacci vitellini
    incompletus
    completus

NOMINA SPECIALIA

Saccus vitellinus
  unilaminaris
  bilaminaris
  trilaminaris[88]
  splanchnopleuricus
  Cavitas vitellina
  Pedunculus vitellinus
    Ductus pedunculi vitellini
Pre-amnion
Amnion
  Somatopleura
  Cavitas amniotica
  Ductus amnioticus
  Liquor amnioticus
Epamnion[89]
  Cavitas epamniotica
Chorion
  Cavitas chorionica (Coeloma extra-embryonicum)
  Trophoblastus
    Conus ectoplacentalis (Preplacenta)
    Cytotrophoblastus

---

[85] Homologues are claws and hooves.
[86] Formerly called the areolar glands of Montgomery.
[87] Derivation from the somatic dermatomes of mammals has not been convincingly established.
[88] This is the condition present in some mammals.
[89] A feature present in ruminants.

Syncytiotrophoblastus
[Syntrophoblastus]
Allantois
  Cavitas allantoica
  Pedunculus allantoicus
  Ductus allantoicus

TYPI PLACENTALES

Indeciduata
Deciduata
Contradeciduata[90]
Villosa
Trabecularis
Labyrinthina
Diffusa
Placentomatosa (Cotyledonaria)
  Placentom [Placentomum]
    Cotyledo (Cotyledo fetalis)
    Caruncula (Cotyledo materna)
Zonaria
Discoidea
Invillosa
Invasculosa
  Placenta vitellina
    unilaminaris
    bilaminaris
    trilaminaris
  Chorionica
  Chorio-amniotica
Vascularis
  Choriovitellina
  Vitellina inversa
    incompletus
    completus
  Chorio-allantoica
Membrana interhaemalis [-hemalis]
  (Limes placentae)[91]

epitheliochorialis
syndesmochorialis
endotheliochorialis
haemochorialis [hemo-]
  haemomonochorialis
  haemodichorialis
  haemotrichorialis
endothelio-endothelialis[92]

ANATOMIA PLACENTAE

Placenta chorio-allantoica
  Pars fetalis
    Zona intima[93]
    Villi chorio-allantoici[94]
      allantoici
      pedunculares
      ramosi
      terminales
      anchorales
    Trabeculae trophoblasticae
    Spatium intervillosum
    Tubuli trophoblastici
    Trophospongium[95]
    Areae absorbentes
      marginales
      disci
      cupulae
      arcus
      vesiculae
    Haematomi [Hematomi]
  Pars materna
    Endometrium
      Decidua basalis
      Glandulae endometriales
      Cupulae endometriales[96]
      Crypti endometriales
      Caruncula (Torus endometrialis)

[90] A placenta which is normally resorbed at parturition, rather than detached; it is claimed for *Talpa europaea*, but is doubtfull in American moles.

[91] Placental types, based on the layers making up the separation membrane: Grosser's classification, and an amplification of it.

[92] Electron microscopy indicates that shrews have a very thin perforate layer of trophoblast separating the maternal from the fetal endothelium. Thus, their placentation may be "physiologically" endothelio-endothelial but "anatomically" endotheliochorial.

[93] An area of intermingled maternal and fetal blood channels: the "labyrinth" (except the trophospongium) of a labyrinthine placenta and the villous zone of a villous placenta.

[94] The villi of all chorio-allantoic placentae consist of a core of vascular allantoic mesoderm and a covering of trophoblast. When they are formed the mesodermal contribution from the chorion is indistinguishable from the allantoic mesoderm.

[95] A trophoblastic zone containing only maternal blood channels.

[96] Now known to be products of trophoblastic cells and to produce chorionic gonadotropins.

MEMBRANAE FETALES HUMANAE

ADNEXA FETALIA

Blastocystis unilaminaris
  Trophoblastus
  Massa cellularis interna
        (Embryoblastus)
Blastocystis bilaminaris
  Omphalopleura bilaminaris
    Trophoblastus
    Hypoblastus extra-embryonicus
  Massa embryonica
    Epiblastus
    Hypoblastus
  Saccus vitellinus primitivus
Blastocystis trilaminaris
  Cellulae amniogenicae [Amnioblasti]
    Amnion primitivum
      Cavitas amniotica
    Discus embryonicus
    Endoderma extra-embryonicum
    Mesoderma extra-embryonicum
      Pedunculus connexionis (Ped.
        corporis)
    Saccus vitellinus primitivus
      Membrana exocoelomica[97]
        Saccus vitellinus proximalis
        Saccus vitellinus distalis
    Cytotrophoblastus
    Syncytiotrophoblastus
      [Syntrophoblastus]
Saccus chorionicus immaturus (Vesicula
      chorionica)[98]
  Cavitas chorionica (Coeloma
      extra-embryonicum)
  Amnion definitivum
    Cavitas amniotica
    Liquor amnioticus
  Saccus vitellinus definitivus
    Pedunculus vitellinus
      Vasa vitellina
      Ductus vitellinus

CHORION
  Trophoblastus

Syncytiotrophoblastus
    (Syntrophoblastus)
  Lacunae trophoblasticae
  Cytotrophoblastus
  Mesoderma chorionicum
Pedunculus allantoicus
  Mesoderma allantoicum
  Vasa allantoica
  Ductus allantoicus
Chorio-allantois
  Chorion frondosum
    Villus primarius
    Villus secundarius
    Villus tertiarius
      peduncularis
      ramosus
      terminalis
      ancoralis
    Spatium intervillosum
    Spatium subchorionicum
      Sinus marginalis
    Cortex trophoblasticus
      Cellula gigantica trophoblastica
        multinuclearis
        uninuclearis
        intravascularis
  Chorion laeve [leve]
Membranae fetales definitivae[99]

PARTES FETALES

AMNION

FUNICULUS UMBILICALIS
  Textus mucoideus connectivus
  Arteria umbilicalis
  Vena umbilicalis
  Ductus sacculi vitellini
  Ductus allantoicus
  Coeloma umbilicale

CHORIO-ALLANTOIS
  Chorion laeve [leve]
    Chorio-amnion
  Chorion frondosum (Discus
      placentalis)[100]

---

[97] Also known as Heuser's membrane.
[98] Three to nine weeks after ovulation.
[99] Ten weeks before parturition.
[100] Also known as the fetal placenta.

Lamina chorionica
Spatium intervillosum
Spatium subchorionicum
Sinus marginalis
Lobus [Cotyledo]
  Villus peduncularis
   Villus ramosus
    Villus ancoralis
    Villus liber
    Lobulus
     Villus terminalis
Syncytiotrophoblastus [Syntrophoblastus]
  Nodi syncytiales

PARAPLACENTA[101]

PARTES MATERNAE

Endometrium basale
 Cellulae deciduales
 Decidua
  basalis
   Carunculae[102]
   Cryptae endometrii
   Septa[103]
   Insulae cellularum
   Glandulae uterinae
  capsularis
  parietalis
 Substantia fibrinoidea

PLACENTA HUMANA

Structura typica
 deciduata
 discoidea
 haemochorialis [hemo-]
  haemodichorialis
  haemomonochorialis
 villosa
Variationes formae
 Placenta accessoria
  annularis [anularis]
  bidiscoidea
  bilobata [bipartita]

circumvallata
diffusa
discoidea
duplex
fenestrata
lobata
lunata
marginata
membranacea
multilobata
multiplex
panduraformis
reflexa
reniformis
sartiginiformis
trilobata [tripartita]
velamentosa
zonaria
Variationes situs
 Placenta dorsalis
 lateralis
 ventralis
 fundalis
 cervicalis (previa)
Variationes fixionis funiculi
 Fixionis centralis
 Fixionis marginalis
 Fixionis velamentosa

DYSMORPHIA

TYPI DYSMORPHIALES

TERATOGENESIS

TERATOLOGY
Errores reproductionis
 Infertilitas
 Sterilitas
 Mors prenatalis
  Abortio
  Resorptio
  Retentio
   cum calcificatione

---

[101] Comprising the *chorion laeve* and *decidua parietalis* where various fetal-maternal exchanges take place.
[102] It is doubted by some that these occur in the human placenta.
[103] Placental septa and cell islands are probably of mixed origin; both trophoblast and decidual contribute.

cum compressione
cum mumificatione
Partus mortuus
Defectio congenitalis
prenatalis
postnatalis
morphologica
simplex
Variatio
Malformatio
Anomalia
multiplex
Syndroma
Monstrum
Tumor monstruosus
functionalis
Defectio metabolica congenitalis
Defectio genetica
Defectio hereditaria
Mutatio
Deletio
Duplicatio
Genum lethale
Genum mutans
Indisjunctio
Inversio
Translocatio
Defectio chromosomalis
Aberratio numerica
Aneuploidea
Aneusomia
Heteroploidea
Hyperploidea
Hypoploidea
Monoploidea
Polyploidea
Monosomia
Polysomia
Triploidea
Trisomia
Abundantia chromosomalis
Deletio chromosomalis
Chromosoma annuliforme [anuliforme]
Satellita
Defectio gametogenetica
promeiotica
meiotica
Aberratio chromosomalis[104]
autosomalis

gonosomalis
Indisjunctio
Fractura chromosomalis
Mosaicismus
Defectio gametica
Defectio fertilizationalis
Gametus immaturus
Gametus senilis
Polyspermia
Zygota corrupta
Defectio implantationis
Implantatio corrupta
Implantatio precervicalis
Implantatio ectopica
abdominalis
primaria
secundaria
ovarica [ovariana]
tubalis
uterina interstitialis
Defectio membranarum fetalium
Defectio amniotica
Adhaesio [Adhesio]
Hydramnion (Polyhydramnion)
Oligohydramnion
Taenia amniotica [Tenia amn.]
Defectio chorionica
Deformitas placentalis[105]
Defectio placentalis
Defectio chorionica nonplacentalis
Defectio funiculi umbilicalis
Funiculus arcuatus
strangulatio
amputatio
Vas anomalum
Defectio embryogenesis
aggregationis
canalisationis
compositionis
conclusionis
conjunctionis
crescentiae
differentiationis
fissionis
migrationis
perforationis
plicationis
retrogressionis
retroplasiae

[104] For karyotype alterations, *see* page E 37.
[105] For a list of variations in shape, *see* page E 29.

separationis
septationis
Embryo defectum
Blastoma
Conceptus abortus
Conceptus corruptus
Deformitas localis
Deformitas multiplex
Dyspraxia
Embryo amorphicum
Gemini conjuncti
Monstrum
Partus mortuus
Tumor monstruosus
Forma abnormalis
Totalis/subtotalis
Fetus amorphicus
Geminus acardiacus
Defectio cordis subtotalis
Defectio cordis totalis
Gemini conjuncti
Gemini symmetrici
paralleli
transversi
Junctio superior
dorsalis (cranialis, craniopagus)
lateralis
ventralis (craniothoracalis, Janus)
Junctio media
xiphoidea
sternalis
thoracica
thoracogastrica
Junctio inferior
dorsalis (clunialis/glutealis)
lateralis (dipygus)
ventralis (coxalis/pelvica)
Gemini asymmetrici (unus
imperfectus)
Hospes
Parasitus
Junctio superior
cranialis parasitica
gnathalis parasitica
Junctio media
thoraco-epigastrica parasitica
abdominalis parasitica
Junctio inferior

pygalis parasitica
Nanus
Nanus achondroplasticus
Nanus atelioticus
Nanus athyroticus
Homunculus
Gigas
Fetus calcificatus
Fetus compressus (Fetus papyraceus)
Fetus inclusus
Deformitas localis
simplex
multiplex

NOMINA TERATOLOGICA
GENERALIA[106]

Aberrans
Aberratio
Abundantia
Absentia
Accessorius
Agenesia
Amorphia
Amputatio
Anaplasia
Anomalia
Aplasia
Astrophia
Atavismus
Ateliosis
Atresia
Atrophia
Bifurcatio
Cataracta
Causa[107]
Choristoma
Coarctatio
Commutatio
Concrescentia (Fusio)
Conjunctio
Constrictio
Crescentia inhibita
Cryptum
Cystis
Dedifferentia
Defectio[108]

[106] Applicable to any structures.
[107] *See* page E 37.
[108] *See* pages E 29–31.

Deficientia
Deformitas
Defectus
Deletio
Dilatatio
Diplogenesis
Dissolutio focalis
Diverticulum
Duplicatio
Dysgenesis
Dysplasia
Dystrophia
Ectasia
Ectopia
Error
Exstrophia
Fibrosis
Fissio
Fissura
Fistula
Gigantismus
Herniatio
Heteroplasia
Heterotopia
Hypermerismus
Hyperplasia
Hypertrophia
Hypomerismus
Hypoplasia
Hyposchisis
Imperforatus
Infantilismus
Involutio
Malformatio
Malpositio
Malrotatio
Metaplasia
Monstrum
Mosaicismus (Tessalatio)
Multilobatio
Nanismus
Necrosis
Neoplasia
Obliteratio
Obstructio
Occlusio
Occultus
Paraplasia
Patentia
Peronia

Persistentia
Pigmentatio
Polycysticus
Polydysplasia
Redundantia
Reduplicatio
Regressio
Repartitio
Retardatio
Retentio
Retroplasia
Rudimentum
Septatio
Stenosis
Syndroma
Subnumerarius
Supernumerarius
Transpositio
Vectio abnormalis
Vestigium

NOMINA TERATOLOGICA SPECIALIA[109]

Ablepharia
Abrachia
Acardia
Acephalia
Acheilia
Acheiria
Achorea
Acrania
Adactylia
Additiones
    Ductus additionalis
    Lobus additionalis
    Organum additionale
    Vas additionale
Aganglionosis
    colonica
    rectalis
Agenesis
    organi
    partis
Aglossia
Agyria
Agnathia
Albinismus

[109] Applicable only to particular structures.

E 32

partialis
totalis
Alopecia
Amastia
Amelia
Amyelia
Anencephalia
Aneurisma
Anhidrosis
Aniridia
Ankyloblepharia
Ankylocheilia
Ankylodactylia
Ankyloglossia
Ankylotia
Anodontia
Anonychyia
Anophthalmia
Anorchismus
Anosmia
Anotia
Anovaria
Anus imperforatus
Aorta coarctata
Aplasia lentis
Apodia
Aprosopia
Arachnodactylia
Arrhinia
Arthrogryposis
Astomia
Atactilia
Athelia
Atrichia
Brachydactylia
Brachyoesophagia [Brachy-esophagia]
Brachymelia
Caecum mobile
Canalis craniopharyngealis
Cataracta
Cebocephalia
Cervix uteri duplex
Chordoma
Cloaca persistens
Coloboma
Cor biloculare
Cor triloculare
    biatriatum
    biventriculare
Costa bifurca
Costa cervicalis
Cranioschisis

Craniosynostosis
Cryptophthalmus
Cryptorchismus
Cyclopia
Cystis
  cervicalis
      branchiogenica
      thyroglossalis
  craniopharyngealis
  dermoides
  pilonidalis
  pre-auricularis
  pulmonaria
  retinalis
  urachalis
Dacryostenosis
Dextrocardia
Dicephalia
Dicheiria
Diglossia
Dignathia
Dimelia
Diphallia
Diplocardia
Diplomyelia
Diprosopia
Dirrhinia
Diverticulum intestinale
      ilei (Meckelii)
Dolichostenomelia
Ductus arteriosus persistens
Dysostosis
  cleidocranialis
  craniofacialis
Dysraphia
Ectocardia
Ectopia
  caeci [ceci]
  cordis
  lentis
  testis
  vesicae urinariae
Ectrodactylia
Ectromelia
Embryoma
Enameloma
Encephalocoelia (Exencephalia)
Epispadia
Eventeratio
Exomphalos
Exostosis
Exstrophia vesicae urinariae

Fibrosis cystica
Fissura abdominalis
Fissura craniospinalis
Fissura facialis obliqua
Fistula
  cervicalis
  pilonidalis
  recto-urethralis
  rectovaginalis
  rectovesicalis
  rectovestibularis
  tracheo-oesophagealis [-eso-]
  umbilicalis
  urachalis
  vesico-uterina
  vesicovaginalis
Foramen interventriculare persistens
Foramen ovale persistens
Foramen primum persistens
Gastroschisis
Glaucoma
Gynaecomastia [Gyne-]
Haemangioma [Hemangioma]
Hemiacardia
Hemicardia
Hemicephalia
Hemicrania
Hemihypertrophia
Hemimelia
Hemivertebra
Hermaphroditismus
  falsus
  verus
Hernia
  diaphragmatica
  inguinalis
  retrocaecalis [-cecalis]
  umbilicalis
Holo-acardia
Hydrencephalia
Hydrocephalia
Hydrocelia
Hygroma cystica
Hymen imperforatus
Hyperostosis
Hypermastia
Hyperonychia
Hyperphalangia (Polyphalangia)
Hypertelorismus
Hypertrichosis
Hypodontia
Hypognathia

Hypohidrosis
Hypophalangia
Hypochromia
Hypospadia
Hypotrichosis
Ichthyosis
Intersexus
Kyphosis
Laevocardia [Levo-]
Lordosis
Macrencephalia
Macrobrachia
Macrocephalia
Macrocheilia
Macrocheiria
Macrodactylia
Macroglossia
Macrognathia
Macromastia
Macromelia
Macrophthalmia
Macroplasia
Macropodia
Macrostomia
Macrotia
Malrotatio intestini
Manus bifurcata
Megacolon
Megagyria
Melanismus
  Naevus iliaris [Nevus]
  Naevus pigmentosus [Nevus]
Membrana pupillaris persistens
Meningocelia
  cranialis
  spinalis
Meningo-encephalocoelia
Meningomyelocoelia
Mesenterium inconjunctum
Micrencephalia
Microbrachia
Microcephalia
Microcheiria
Microdactylia
Microglossia
Micrognathia
Microgyria
Micromastia
Micromelia
Microphthalmia
Microplasia
Microstomia

Microthelia
Microtia
Mongolismus
Monopodia
Mosaicismus
Myelocoelia
Myeloschisis
Naevus [Nevus]
  pigmentosus
  vascularis
Nephroblastoma
Neuroblastoma
Notomelia
Omphalocelia
Onychodystrophia
Osteogenesis imperfecta
Otocephalia
Ovotestis
Oxycephalia
Pachycephalia
Pachyglossia
Pachygyria
Pachyonychia
Pachysomia
Palatum fissum (Uranoschisis)
Pancreas annulare [anulare]
Peromelia
Phocomelia
Plagiocephalia
Polycoria
Polydactylia
Polygyria
Polymastia
Polymelia
Polymerismus
Polyodontia (Hyperdontia)
Polyonychia
Polyorchismus
Polyotia
Polyovaria
Polyphyodontia
Polysomia
Polythelia (Hyperthelia)
Rachischisis
Rachitis fetalis
Ren glomeratus
Ren lobatus
Ren pelvicus
Ren polycysticus
Ren sigmoideus
Ren unguliformis
Scaphocephalia

Schistocephalia
Schistocheilia
Schistocheiria
Schistocoelia
Schistocrania
Schistoglossia
Schistognathia
Schistomelia
Schistomyelia
Schistopodia
Schistoprosopia
Schistopyelia
Schistosomia
Schistosternia
Scoliosis
Sinus
  branchiogenicus
  coccygealis
  dermalis
  pilonidalis
  preauricularis
  urachalis
Sirenomelia
Situs inversus visceralis
  partialis
  totalis
Spina bifida
  aperta
  occulta
Symmelia
Sympodia
Syncheilia
Syndactylia
Syndroma
  Arnold-Chiarii
  Downii
  Eisenmengerii
  Fallottii (Tetralogia)
  Frölichii
  Klinefelterii
  Klippel-Feilii
  Lawrence-Biedlii
  Marfanii
  Turnerii
  et cetera
Synotia
Synostosis
Talipes
Talipomanus
Teratoma
Tribrachia
Tricephalia

NOMINA EMBRYOLOGICA

Tripodia
Uranoschisis [Palatum fissum]
Urachus persistens
Uterus bicornis
Uterus duplex
Uterus infantilis
Uterus septus (bipartus)
Uterus unicornis
Ventriculus thoracicus
Volvulus

NOMINA DYSFUNCTIONIS[110]

Amyotonia
Deficientia
  functionalis
  histogenetica
  organogenetica
  reactionis
  secretoria
  sensilis
    Amaurosis
    Anodynia
    Anosmia
    Atactilia
  stimulus
  synthesis
    Defectus metabolicus congenitalis
      Acidi aminoici (Alkaptonuria)
      Acidi nucleici (Purinaemia)
      [-emia]
      Carbohydrati (Galactosaemia)
      [-emia]
      Mineralium (Asideria)
      Pigmenti (Methaemoglobinaemia)
      [-emia]
      Steroideorum (Syndroma
        adrenogenitale)
  tonalis nervi
Dystrophia (Lipodystrophia intestinalis)
Incompatibilitas immunalis (Anaemia
  haemolytica)
Pseudohypertrophia muscularis

Conjunctio (Pancreas annulare [anulare])
Crescentia abnormalis
  Organismus totalis
    Asymmetria (Hypertrophia
      unilateralis)
    Amorphia (Fetus amorphus)
    Hypertrophia symmetrica
      (Macrosomia)
    Deficientia (Microsomia; Nanismus)
      hormonalis
        pituitarius (Nanus pituitarius)
        thyroideus (Nanus cretinicus)
      vitaminalis (Achondroplasia
        fetalis)
    Defectus plasmaticus (Aprosopia)
  Pars localis/Organum locale
    Deficientia
      Agenesis (Anephria)
      Atresia (Meatus acusticus solidus)
      Hypoplasia (Uterus infantilis)
      Defectus canalisationis (Ductus
        lacrimalis solidus)
      Defectus occlusionis
        Apertura persistens (Fistula
          urachalis)
        Fissura persistens (Spina bifida)
        Patentia persistens (Ductus
          arteriosus)
      Defectus separationis (Cyclopia)
      Defectus septationis
        intracardiacus (Cor triloculare)
        intracoelomicus (Hiatus
          phrenicus)
    Abundantia
      Gigantismus localis
        Hyperplasia
          totalis (Macrodactylia)
          partialis (Sphincter pylori)
        Hypertrophia (Megalocardia)
      Supernumerarius
        Multiplicatio organi (Polymastia)
        Superlobatio (Lobus hepaticus
          accessorius)
        Superpartitio (Polydactylia)

STATUS ABNORMALIS

Agenesia (Amelia)
Atavismus (Lobus azygos pulmonis)

ABNORMALITAS INDUCTIONIS

Absentia (Aphakia)
Deficientia (Tumor monstruosus)

---

[110] Most entries in the following lists will be illustrated by an example placed within parentheses.

Impedimentum (Phenylketonuria)

ABNORMALITAS ORGANORUM

Defectio
  conjunctionis (Palatum fissum)
  migrationis
    Deficentia
      cellularum (Megacolon
        aganglionicum)
      organi (Cryptorchismus)
    Abundantia (Ovarium degressum)
  perforationis (Anus imperforatus)
  plicationis (Schistomyelia)
  retroplastica (Vena cava duplex)
  synthesis (Tyrosinosis)
Duplicatio
  Partis localis
    partialis (Ureter bifurcatus)
    totalis (Ureter duplex)
  Organum (Gonada supernumeraria)
  Organismus totalis
    Polyembryonia
    Corpora conjuncta[111]
Ectopia
  hernialis (Meningocoelia)
  inversionalis (Dextrocardia)
  originalis (Dens palatinus)
  translocationalis (Intestinum
    nonrotatum)
Exstrophia (Vesica urinaria eversa)
Hermaphroditismus
  falsus
  verus
Inversio
  partialis (aorticopulmonaria)
  totalis (thoraco-abdominalis)
Persistentia
  Atresiae ad tempus (Duodenum
    imperforatum)
  Formae fetales (Ren lobatus)
Syndroma (Downii)

ABNORMALITAS TEXTUS

Abundantia
  dermalis (Naevus [Nevus] vasculosus)
  epidermalis

Stratificatio (Ichthyosis)
Pigmentatio (Melanismus)
Naevus [Nevus] pigmentosus
Excrescentia cartilaginea (Nodulus
  cartilagineus)
Excrescentia ossea (Exostosis)
Neoplasia
  Textus neuralis (Neuroblastoma)
  Textus notochordalis (Chordoma)
  Textus renalis (Nephroblastoma)
Ametastasis (Megacolon)
Heteroplasia (Cartilago renalis)
Hypofunctio
  metabolica (Diabetes)
  somatica (Stenosis pylori)
Deficientia
  Agenesia
    cellularis (Cerebrum acallosale)
    intracellularis [intracellum]
      (Albinismus)
  Dysplasia
    ectodermalis
      epidermalis (Ichthyosis)
      neuralis (Imbecillitas amaurotica)
    multiplex (Lipochondrodysplasia)
  Ossea
    Achondroplasia
    Osteogenesis imperfecta
Retroplasia (Chondrocathodia)

CAUSA TERATOGENICA

Genetica
  Karyotypus modificatus
    Aberratio numeri
      Gonosomia
        Monosomia
          (Syndroma Turnerii)
        Trisomia (Syndroma
          Klinefelterii)
      Autosomia
        Monosomia (Monosomia 22)
        Trisomia (Trisomia 21,
          syndroma Downii)
        Mosaicismus (Syndroma
          Edwardsii)
    Aberratio morphologica
      Deletio (Syndroma vagitus felini)
      Duplicatio

[111] Types of conjoined twins are listed on page E 31.

Inversio

Isochromosoma

Translocatio (Syndroma Downii)

Mutatio genorum

Genum autosomale

dominans (Ren polycysticus)

recessivum (Haemophilia) [Hemo-]

Genum gonosomale (Chromosomale sexuale)

dominans (Rachitis antivitaminica)

recessivum (Haemophilia) [Hemo-]

Functionalis

Deficientia

Stimulus

Reactionis

histogenetica

organogenetica

Humoralis

Deficientia (Cretinismus)

Abundantia (Masculinisatio)

Incompatabilita immunalis (Anaemia haemolytica) [hemo-]

Infectiosa (Virus)

Ignota

Vicinalis

Chemica

accidentalis (Pollutio)

therapeutica (Thalidomidum)

mechanica (Trauma)

nutritionalis (Hypervitaminosis)

Physica (Radiatio)